The Science Of Physics

Proof That God Exists

Michael Mathiesen

A human being is part of a whole called by us - "Universe."

Albert Einstein

Copyright 2019 - © Michael Mathiesen
Cover by Michael Mathiesen

Table of Contents

<u>Preface</u>

I must tell you first that I am not a Scientist in that I am not now nor have ever been employed by an educational or scientific organization. This has never been part of my career. Instead, I have been an independent thinker all of my life as far back as I can remember but in my independent research, I feel that I have been freer to investigate the questions that truly need to be answered, rather than minor insignificant questions. And the biggest question I've had throughout my life is how events in my life are ordered, by whom are they so ordered and to what purpose are they so ordered? Are these indications of some kind of universal laws? I think that you might agree by the time you finish this book that I have some answers about that.

The 'Hand Of God' Nebula - pulsar B1509-58.

Throughout my life, I always felt the influence of a hidden hand, some guiding and loving force that was always there like a shield protecting me from evil and always leading to the light. I believe that in this book, I finally layout some of the best answers to these eternal questions in a way that can be understood by anyone with an active brain. In this book, we cover some of the most intimate and delicate problems in Quantum Physics, Cosmology and the Biological Sciences as well since we will end with one of the most challenging problems in genetics we will ever face as a species.

And this book also gives us a novel way to look most objectively at this particular moment in our own Evolution where we could be on the edge of total extinction of all life on the Earth, including ourselves, or we may be on the edge of a completely new way of managing our society and our planet in ways that preserve and sustain them in a new economic and political system that is fair and equitable for all life on the Earth. But, this must include you for it to work, I'm afraid. Not to worry, I have faith in you.

Although, these are big concepts, the Science I present is easy enough for anyone with a High School education to understand and follow. In all of my writings, I always attempt to reduce things down to the simplest of terms so that I can understand and uncover the larger perspectives and present them to my readers in this highest and best form.

At the end of the book, I hope you will conclude that I've

done a decent job at making these extremely important issues of life and death jump off the page and command your attention as well as your comprehension and acceptance as the best interpretation of the evidence.

Introduction

You've probably heard people say to you or perhaps you said to someone else that "Everything happens for a reason!"

You usually hear this old adage expressed when there is a sad, disappointing or heartbreaking event in your life or the life of a friend or family member, especially just after an unexpected death. I am the type of person who tests just about everything that he learns and starting the day I first heard this overly used phrase I tried to figure out if this is true for everyone everywhere in the universe as is implied. Is this an immutable law in regards to the evolution of the universe and especially all of us who live here? I've explored this theory in many of my previous books, but in this book, I believe I have solved that question and put it to bed for all time.

Have you ever thought about that little voice that speaks to you constantly from deep inside your head? Who is that? Is it you? Or is it someone you would like to be? Is it ever turned off? Do you remember when you first heard it talking to you, making plans for your current day's activities, telling you how to think about yesterday's activities, or helping you plan for to-morrow?

When you were first born, you had no voice. In your youngest iteration, your inner voice couldn't really talk to you until you learned your native tongue, your language. Until you picked up on words, you had to be thinking like a dog or a cat must think, with images, instead of words. Did you ever think that you would always have that voice? That somehow, your voice would never die? When you finally do get old and pass away from this planet, does your voice go along with you - you wonder?

It's the same thing as thinking about your favorite music. Does music go away when you stop hearing it? No, I don't think so. Music stays where it always hangs out because the next time you need it, it's there again, just like your voice, waiting for you to listen to it. The greatest music composers know this and that's why they are able to tap this wonderful amazing form of energy and share it with the rest of us when-ever they want.

So, you may want to ask - 'Where does music hang out?' You may want to know, 'Where does that voice inside my head go, when I don't need it?' And, it's OK to ask these questions and more questions like these as you go along in my book because in this book, we're going to be studying this kind of thing, the scientific and spiritual laws of Physics because I believe that the Scientific explanation for things that we read about and learn about in school, in college, from other scientists and the spiritual explanations that we all feel sometimes and learn about in Church are actually the same thing.

The stories and fables they teach us in Church, of course, are very primitive attempts to define what God would look or sound like, but too many times, these depictions of God are way off of any normal type of common sense. The Bible Story in the opening chapters of what they call 'The Holy Scriptures', for example, where God tells Abraham to slit the throat of his own son just to see if he, Abraham, is truly faithful to God. This kind of thing, God would never tell any of us to do because there is no God in my mind that would be so insecure that he needed to find out if a single human being truly thought of him as the God of all Gods. It's preposterous and shameful that many millions of people over the years have put their faith in this kind of dogma, meaning stories that have been repeated over and over so many times that it's OK to think of them as true even though they are far far away from the truth..

But, even when confronted with the Church dogma from this and other badly chosen fairy tales, I still believe, have always believed, and will always believe that there is a God somewhere out there in the Cosmos and I base this faith and trust in this truth, not from silly horse nonsense that others have put down on the page, but from my own observations of the largest things in the universe, the stars and galaxies in the heavens, and from the smallest things in the universe, the sub-atomic particles, as well as from the events of my own life.

When I have become aware of the greatest scientific discoveries, things we take for granted today as Truth, I can easily see the hand of the Creator of these things. None of the things that I will show you in this book could possibly have happened without an architect of the building blocks of the universe because the Creator's finger prints and/or footprints are visible to the naked eye or to one's ears or any of our other senses.

As one overwhelming and easily understood example of this let's talk about the way that our home planet the Earth revolves around the Sun so that it always remains in what we have now called the Goldy-Locks Zone, the perfect distance, give or take a few miles, from which a planetary body like ours can get just enough heat, light and where water can exist as a liquid so that life becomes a distinct possibility, and not only a possibility but can be guaranteed a position in Space/Time that could last as long or longer than the sun itself.

Some people, even today, will argue that this scientific fact is a mere random coincidence. They will point to the billions of other planets that we are discovering all around our galaxy where the distance from the planet to their own star is too far and would therefore be too cold for life or they would be too close whereupon they would be too hot to support life. They would point to the other planets in our own solar system like Venus, Mercury, Mars, Jupiter, Pluto etc. where we find no life because these other planets are all either too close or too far from the Sun, and therefore, the Earth's beneficial distance from the Sun is merely random chance, not a miracle that causes us to see the workings of a Creator in this Cosmic scheme of things.

But, it's not just this one piece of evidence that informs my opinion on the subject. There are countless other things about how the universe works for us as living beings that makes me tally up the score either FOR or AGAINST the argument that God exists and I have to say that so far, the score is 'That God Exists' =1,019 For and =2 Against the proposition. That score is good enough for me. The 2 bits of evidence that there is no God, I am going to save for the end of this book. You can probably guess what they are because they are big ones, big arguments against there being a God and I wrestle with those two BIG ONES in my head quite often and mostly when I'm feeling the weight of these two on my own life. But, in the final analysis, I usually put them down to the fact that we are judging these things from the wrong point of view and when we unite with our Creator, He or She, will make a point of explaining this little anomaly in the data to all of our satisfaction.

Sadly, we all have Ego's you see and this part of our Consciousness makes parts of the universe be taken in to our senses in a fog of some kind and some things are positioned in this blurring of life to look far more menacing to our Egos than they really are. It's hard to feel our own Egos in these situations because that's the part of our brain that others control, or at least greatly influence, and so we may not always be rational. In fact, I know we're not always rational and live in a highly subjective state of awareness part of the time.

To put an exclamation mark (!) on this type of idea, I am inserting here my favorite photograph and what I believe is the most important photograph ever taken.

It's taken from our Space probe Voyager in 2012 as it left our Solar System, the first man-made object to ever do so. I wonder how many of you can tell what this is all about.

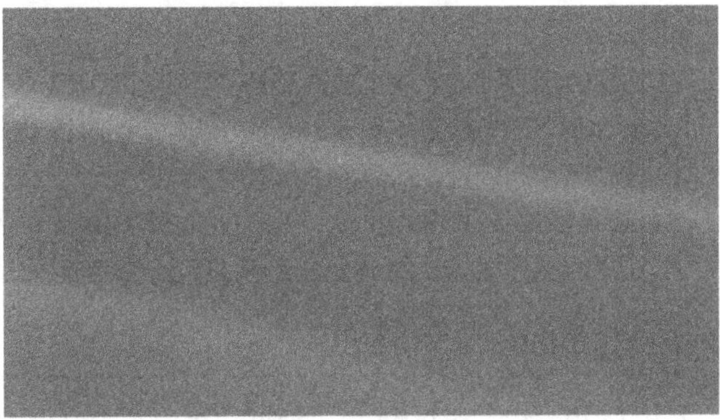

As the Space Craft Voyager left our Solar System, Mission Director Carl Sagan had a signal sent to the space craft to turn around and snap one last photograph. Not knowing what would come out of it, the NASA team waited several hours for the data that would be assembled into this picture above.

At first, they thought that the tiny white spec you see in the yellow beam of light in the Top-Middle of the picture was a grain of sand or an imperfection in the development of the photo. One of the NASA people tried to wipe it off the page. It wouldn't budge. So, upon closer examination, it became clear that the white dot in the Top-Middle of the picture is our own planet Earth, the originating point in Space from whence the little space craft was launched thirty-five years previously.

The finger of orangy-yellowish light that the Earth is floating in is merely that, a finger of light emanating from the Sun, about 90 million miles to the leftward orientation of this picture and very graciously pointing out for everyone to see just how small and insignificant a place our planet takes in support of all our words, deeds and actions.

A person can look at this photo very coldly and objectively and see only what is shown and make no major interpretations of this event other than the astounding fact that the equipment on board Voyager was still working in this completely hostile area in space and after so many years past everyone's life expectancy of the intricate hardware involved.

Or a person can learn that this picture was never planned in all of the thousands of hours of designing the space craft's mission. They would have their hands full orienting the space craft to snap millions of photos of Mars, Jupiter, Saturn, the moons of Saturn, Uranus and Pluto as it flew past at about fifty-thousand miles per hour. No one ever thought to turn the cameras back for one final shot from this distance where they knew that nothing would be there worthy enough to photograph.

In fact, it was merely the suggestion of one of Carl Sagan's assistants to send one last command to the space ship to turn around and take a random photograph like this. The brass at NASA had rejected the idea thinking that nothing could come of a very expensive, multi-million dollar maneuver. And besides, they told Director Sagan, the project had been shut-down. Everyone had gone home.

Carl Sagan persisted however, because it just seemed like such a cool idea and eventually he was able to convince his bosses to call several project engineers back to their command offices at Jet Propulsion Labs and get to work to send one last signal to Voyager.

And so it was done. It amazed everyone at JPL that Voyager still had enough juice in its batteries to not only listen to their signal but also to obey and turn itself around, focus it's camera in the exact opposite orientation to where it was going and snap one last photo and transmit the data back to the Earth.

Whenever I assess all of these points of data, I come to the only conclusion that someone like myself, and I hope you too can make.

The little Voyager space-craft had one last job to do. It somehow made this known to its creators. They took up the challenge and allowed Voyager to turn around and wave good-bye to us, the creatures who had made it, and in doing so, would memorialize us and point out to us exactly how rare and precious an element we are in the overall scheme of things.

Now, I have to say it - Compare this actual event that is abundantly recorded in digital form that will last for millions of years to the story of Abraham when God told him to kill his only son. Which of these events would you trust as proof that God exists?

And if the Human Race goes extinct someday and no other living beings are able to contact us, they may someday bump into this little space craft that we constructed and now flying deep in inter-stellar space. They will take it apart and learn how to play the music and videos that we placed on a golden record filled with images of our own lives and many of the life forms whom we are honored to share this planet with. They will hear music from Bach, Beethoven, the Beatles and Chuck Berry. They will see images of babies, flowers, exotic fish, animals running wild in the jungle, on the savanna, Eagles and other majestic birds floating majestically in the sky. They will even be shown a map of how to find our planet.

When they arrive, they may find that we have destroyed everything here and that there are no wondrous animals and plants adorning this Earth any longer. There may be nothing but another burned out and lifeless planets left for them to see.

And if they are good creatures, they will memorialize us and maybe even feel sorry for us. They will at least, from the records and data on board Voyager, know what we looked like, what we have striven for most arduously, the kinds of things that made us happy and even how much we loved each other.

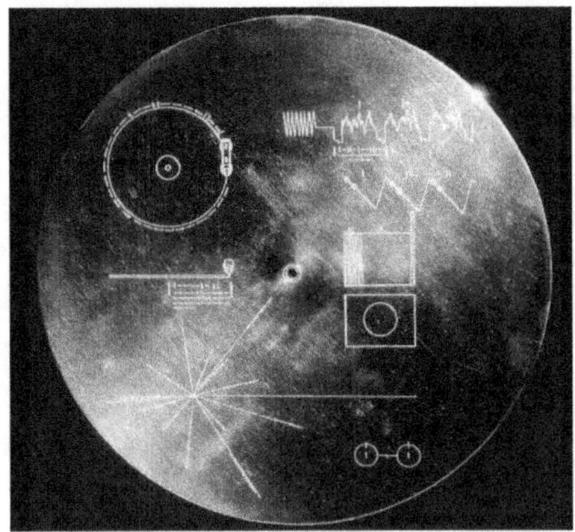

But, they may never be able to fully understand how we lacked the ability to save ourselves.

Chapter One
- The Standard Model

The following graphic is a chart of all the major groupings of the known energies, forces and particles that make up our universe. It's called the 'Standard Model' of the smallest segments of the universe, the atomic and sub-atomic world and all of our latest evidence from extremely rigorous experimentation all over the world for centuries supports this model and makes it the greatest and most widely accepted standard of 'Truth' as far as we have the ability to know it.

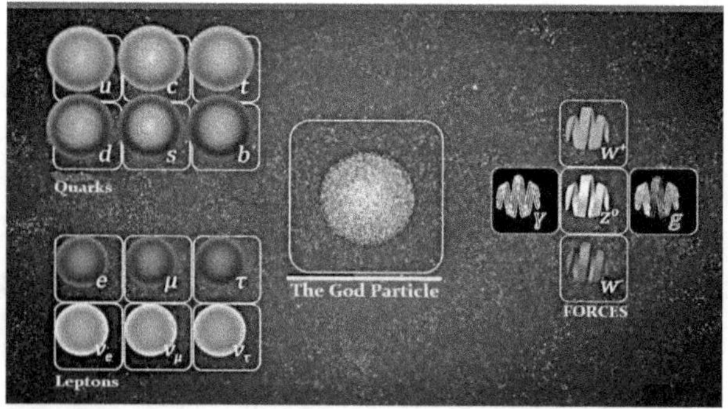

This is a chart of everything we know that makes up all parts of the 'ATOM'. We all remember our Elementary School Science teacher showing us the model of the Atom with at least one Proton and a Neutron making up the nucleus and the Electron making a spherical shaped orbit around the nucleus.

The Top Left Box includes all of the known quarks, the smallest and most fundamental building blocks of every kind of matter that we know. As you can see there are SIX of them. As far as our research has taken us so far, the Quarks are the smallest and most indivisible of all of the sub-atomic particles. Two thousand years ago, the Greeks told us that the 'Atoms' would be the smallest and most indivisible object in the universe from which all things are composed. But, as we know Science never sleeps and we now know or believe that Quarks take this role.

Protons are made out of these things called Quarks, at least three of them and sometimes hundreds of them that are floating around inside the Proton and their energy as they bump around in this tiniest of space is what makes up the 'mass' of the Proton.

The six Quarks are named: 'Up', 'Down', 'Charm', 'Strange', 'Top' and 'Bottom'. Their names have no basis in anything other than a convenient way to label them. You can think of the 'Up' Quark takes up a higher position within the Proton and the 'Down' Quark is usually sitting below the 'Up'. The rest of the names of the Quarks are usually stated as 'Flavors' even though no one has ever or will ever be able to taste anything this small.

Indeed, you may think of this explanation of the Physical world as 'Strange' and you would be right because one of the Quarks is indeed labeled as such, even though you could call them all 'Strange'.

The bottom left box is filled with the Leptons. There are six different types of leptons. These include the Electron, the Muon, and Tau particles, as well as their associated neutrinos (i.e. electron neutrino, muon neutrino, and tau neutrino). All Leptons have a negative charge and a distinct mass, whereas their neutrinos have a neutral charge.

We are most familiar with the activities of one Lepton, the Electron because we use 'Electricity' the flow of Electrons in our daily lives. All Leptons have a negative charge and we're familiar with that because if we've ever been 'Shocked' by touching an electrical cable, we have experienced this negative charge and what is the result of the flow of negatively charged electrons around and through our simplest and most complex circuitry.

You may even be reading these words through or over an 'Electron' connection between your own computer and the Internet of computers. The Muons, the Tau Particles are probably capable of generating their own kind of electricity because Muons have a mass of .12 Gev (GigaElectronVolts) while Electrons have a mass of only 0.0005 Gev, making the Muons approximately 200 times more powerful (more mass = more energy) than the Electrons. And, even more powerful, the Tau particles have a mass of 1.777 Gev or about 4,000 times more powerful than the Electron.

This leads me to predict that someday, Humans may be able to use this type of sub-atomic energy and get thousands of times more energy out of this part of the universe than we are getting right now out of the Electron energies. The only thing we have to wait for is the scientific knowledge to determine how to do that. Somewhere there may be another Thomas Edison out there who is will be lucky enough to find a practical way to do it. The point is, we know very little about this level of the universe after only discovering it for a few hundred years, but we're getting there at a quickening pace.

The other interesting thing about Leptons is that they can change their flavor at will. In other words, an Electron can decay into a Muon and a Muon can change into a Tau Particle. Protons can morph into Neutrons and Neutrons can morph into Protons. It seems to this author that this tiniest realm or 'Medium' allows someone or something, like a great artist manipulate all of this stuff to suit his or her artistic whims and that this aspect of the universe is very much like the clay that a sculptor may use, or the paint that an artist may use or more accurately the musical notation that a great composer may use to depict his or her interpretation of the world.

Put another way, we have to wonder why we have all of these basic particles and laws of interaction when it appears that random chance (or not so random chance) can alter these particles by changing the rules at will?' The universe, it seems to me is made up by the greatest Realist painter, because this artistry is how we achieve the 'Reality' we live in, rather than a complete and total random series of events.

Stranger still, all of these subatomic particles also have an equal but opposite particle called 'Anti-Matter'. I have a previous book, Science Fiction, with the Title 'Anti-Matter' because it occurs to me that the universe we live in being controlled by the interaction of 'Matter' and 'Anti-Matter'. We simply don't know much about Anti-Matter except that it does in fact exist and we can even create it in the laboratory. But, sadly, whenever 'Matter' - any of these particles we're discussing - comes into contact with a particle of the opposite universe, 'Anti-Matter', the two particles totally annihilate each other while giving off a small puff of energy in the form of photons. In my previous work, I speculated on where all of this matter goes and what impact this reaction may have on the events in our lives.

The Plot Thickens:

The box on the right of these other two in the chart of the Standard Model above composes the known forces. There only four main forces in the Universe and we are familiar with one of them - Gravity. The other three:

Electro-magnetism (Electricity), and the Strong Force and the Weak Force. The Strong Force is so-called because it is the strongest force we have discovered, but it plays its role only in the smallest of distances.

The Strong Force is the force that holds the Protons together in a very tight concentrated ball in the middle or the 'Nucleus' of the atom. You may also recall that Protons carry a positive charge and in most atoms there are more than one Proton sitting close to their friends. But, if two charges are the same, they will repel one another. And so the atoms would all explode apart and there would be no universe as we know it if it were not for the Strong Force that holds the protons together, overcoming their tendency to explosively break apart. This is why it is called the 'Strong Force'. It is indeed very strong - thankfully.

The Weak Force is the force that is the opposite of the Strong Force. It allows the sub-atomic particles to decay or change into something else. It is the force that we have tamed in Nuclear Power plants because these machines gain their energy to generate power to our homes through the decay of the uranium atoms, mainly the neutrons. If it were not for the 'Weak Force' - oddly enough extremely powerful for our use - we would not get any power to power our homes and the Sun would not be able to keep us warm because the power of the sun emanates from the decay of Hydrogen atoms into Helium atoms.

Even weaker than the Weak Force, however, is Gravity. And Gravity is not included in the Standard Model of sub-atomic or Quantum Physics because Gravity is so weak it has simply never been measured inside the Atom. In other words, there is no evidence that any of the inter-actions within the atom are impacted by Gravity and therefore, it simply does not fit inside this generally accepted understanding, the Standard Model of the universe and it therefore - so far is kept apart.

You may be old enough to remember how our Elementary School Science teachers told us that the Earth has a large mass and the Sun's mass is even greater and so the Earth remains in its orbit around the Sun because of the attraction of the two masses, that of the Earth and that of the Sun for each other.

Unfortunately, that is no longer the Truest definition of this force.

Instead, today Gravity is better understood in terms of Einstein's great theory of Relativity which we can all remember describes the universe in its largest of events out there in Space. Sir Isaac Newton taught us that the interaction of all the planets and stars can be explained in terms of Gravitational attraction but Einstein amended this rule and has described this attraction of any two planetary bodies as a 'Warp' in the FABRIC of Space/Time. This means that the Earth flies around in its orbit around the Sun because the Gravity of the Earth and the Sun working in tandem has created a groove in the fabric of Space/Time, the stuff that lies between the Earth and the Sun and the Earth merely falls into that groove. The moon falls into a smaller groove in the fabric of Space/Time which is created by the two masses of the Earth and the Moon.

This means that there is another much larger groove in the fabric of Space/Time that is created by the mass of the entire galaxy in relation to the mass of the Sun, and so the Sun simply falls into this groove, carrying all of the planets with it, and makes its silent journey orbiting the center of our Milky Way Galaxy.

These grooves in the Fabric of Space/Time must exist everywhere in relationship to everything else because we know that even galaxies are lying inside a groove that will eventually cause them to crash into each other and then gradually blend seamlessly into a larger galaxy.

We also know that all of these things that I've described have been going on since the beginning of Time as far as we can measure it.

And now we come to the discussion of the little thing in the middle of the Chart of the Standard Model.

The God Particle:

Gravity is not included inside the Standard Model simply because no one - not even Einstein - has been able to connect Gravity to any of the sub-atomic forces or particles. Einstein was not able to connect the dots merely because he was not able to think outside the box. As brilliant a genius as he was in finding out that matter and energy are one in the same thing, he couldn't complete the mathematical connection between the largest force in the universe with the smallest forces in the universe and the reason for this is that the forces in the universe are counter-intuitive.

The largest or strongest force in the universe is The Strong Force, because it holds together the protons who want to fly apart and keeps them cemented together creating the nucleus of the atom. But this strongest of forces only takes hold of things over the smallest distances in Space/Time. The weakest of the four known forces - Gravity - takes hold over the largest areas of Space/Time. And that - even though known to Einstein - never registered in his mind in the twisted sense of reality that it represents.

One of the strangest discoveries of my life was the day

that I learned that Gravity waves can fly past the Earth from the opposite side of the universe. We don't have any instrumentation currently that can pick up on these gravity waves, but we know this happens because of a recent device being put into operation known as LIGO, the Large Interferometer Gravity wave Observatory. This is the most sensitive instrument for observation of the heavens ever created and the minute they turned it on in 2015, it recorded gravity waves from 1.3 billion light years away.

Gravity waves from this distance are not going to make the tides go up or down. They're far too weak to make any changes to the Earth's orbit around the sun or to any perceivable changes to any orbits of any planets in this or any solar system. However, they are strong enough to stretch the fabric of Space/Time, even though less than a trillionth of a trillionth of a millimeter. We now know, from LIGO and many other similar observations made by the several other LIGO's that have been put into operation since - that even gravity waves from this far away have an impact on the Space/Time all around us.

But Albert Einstein, since he died in 1957 was never aware of any experiments that would give us a direct observation of gravity waves, which took place in 2015 and he was never aware of the proof that the Higg's boson or The God Particle until the scientists at CERN discovered it in 2012.

Now, if you turn back to the first page of this chapter and look at the Chart of the Standard Model of sub-atomic physics, you see where I have notated 'The God Particle' in the middle

of all the other known particles and forces. It's important to
note that this particle was missing from this chart until 2012,
when it was proven to exist. And we now know that it is a central player, and even more so, a Keystone - literally - in the architecture of everything else. Until that time, it had only been
theorized to exist by several Physicists some 50 years previously, some of them, one a professor Peter Higgs of Scotland's
University of Edinburgh.

The scientists at the time had all concluded that there had
to be a particle that would carry the property of 'mass' all over
the place and impose it upon all of the particles, since all of
them, other than photons and gluons have enough mass for us
to measure. It was experimentation done by the Large Hadron
Collider and announced to the world on July 4th 2012 that
made the Higgs boson an official member of the Standard
Model of all things and placed it - justifiably in the center of
the model. It took it's rightful place in the center of the chart
because without it none of the other particles can exist.

And this is why the Higgs boson achieves the nickname of
the of 'The God Particle' because it would have to be the first
particle that God created in order for the rest of the universe to
evolve in the way that it has. Since this name is extremely de-
scriptive of the role that it plays in the universe, it is my con-
tention that it will stick for thousands of years and not much
longer be remembered as merely 'The Higgs' as most scientists
want to do today.

Most scientists don't like to use the word 'God' because
the Creative Force that comes before all other known things in

the universe cannot be tested and therefore cannot be proven to exist - and this is the essence of the Scientific Method to be sure, which I greatly respect.

However, just because we cannot have a test for the existence of a creative force, doesn't mean that it cannot exist. In fact, to my way of thinking, the evidence is overwhelming that God exists and this is the nature of this book. It's a proof that I believe will stand the test of time - and therefore it will be tested and someday even proven if only in the minds of the vast majority of people on the Earth and/or elsewhere. Just because a few Scientists are unable to accept this truth doesn't mean that these few are right. Sometimes when something walks like a duck - it's a duck.

The next thing to know about the God Particle is how it gives mass to every other particle in the universe and how we know it is the messenger of God.

First of all, whenever you read about a particle in the field of Physics, you must remember that a particle at least at the sub-atomic level is really just a point in Space/Time. It does not exist all by itself. All sub-atomic particles exist in a kind of field of probability. In other words, we can tell that electrons exists because we can actually see their impacts on the rest of the atom and we know how to manipulate them by using a magnetic field to send them down the wires and into your homes and offices. So, all sub-atomic particles exist in a kind of flux, that is to say they are really a measurement of a particular place in Space/Time that is all part of a larger field. So, electrons only exist because they are given their existence, their mass, by the God Particle Field, or the Higgs Field.

So, the God Particle Field is everywhere that we can see as part of our universe. How do we know that? Because everywhere we look in the universe we see the same stuff of electrons, protons, neutrons, quarks, etc. And we know now, since the discovery of the God Particle Field that everything has to receive its mass from the God Particle Field. But, how does it happen? It's just like any other field. Where an Electron can exist, there must be an Electron Field, or the Force of Electromagnetism. This makes an electron from out of nowhere, nowhere but the field. If we're lucky a Proton is formed up in this same way nearby and so the Electron and the Proton can come together more easily and thereby make up the 'Atoms'. Similar forces, we'll talk about later combine to make the 'Atoms' join forces and become the molecules. Some of these molecules join forces to become proteins. Some of these molecules join forces to become the DNA molecule which will become the basis of LIFE on the Earth and if I'm correct in my theory of life, DNA will be discovered as the basis of life all over the universe, not just on the Earth.

But, since the God Particle Field is everywhere, every known particle in the universe can be induced out of what is described as a 'Condensate' or a fancy way of saying a 'Soup'. The God Particle Field soup simply creates anything in the size and type of a sub-atomic particle at any time and at any place. It does this by simply becoming 'Excited' at the point in Space/Time where it wants to give birth to such an object.

Imagine, if you will, the 'Excitement' that must exist as

you create the universe with one tiny bit after another. In some places you need lots of bits, and in other places you don't want any bits. Instead, you may want only darkness and in other places you may want to create huge sources of light.

And this is how the Bible got it fairly correct when it says in the book of Genesis - "And God looked upon the darkness and said, 'Let there be light'. And then there was light all across the land."

Actually, if God says anything at all to himself, he might be better described as saying - "And God looked upon the darkness and said, 'Let there be the God Particle Field, wherein I can plant all of the seeds of a much larger realm. And, there was a God Particle Field that came out of the nothingness and they shall call it The Big Bang.'"

And, who is this 'They' that God is predicting? It's us, of course, the human beings and of all other life forms that we know since we have evolved from earlier forms of life here on this planet.

The main point here is that all of the most basic building blocks of the universe are made seemingly at random. In fact, for millions of years, human beings have looked at this kind of Creation as just random. The Bible for example doesn't explain why God made the world, just that he did it one day as a kind of random act. All holy scripture from all the other age-old religions also imply the same kind of motivation for God creating the world. He just did and we're all stuck with what he did and we just have to 'live with it'.

Today, however, because of our Scientific achievements, some of which I have described already, we are able to view almost the entire universe and we can readily see that there are millions of words we might use to describe this magnificent universe - but 'Randomness' is not one of them.

Everything we can see is the ultimate definition of order and beauty.

Most of us have seen the magnificent panorama of stars that make up our night sky.

What we see in the night sky is one spiral arm of our Milky way galaxy. There are several arms and we are on one of the outside arms, nowhere near the center of the galaxy.

That could be a good thing because we know that at the center of every galaxy is a Black Hole.

The Milky Way Galaxy

Our own galaxy is filled with about four hundred billion suns like our own, some bigger, some smaller. Most of these suns, we believe, have at least one planet not unlike our own. Also notice in the bottom of this picture another galaxy floating around nearby. It's the Andromeda Galaxy that is on a collision course with our own and due to collide into our own galaxy in about a billion years.

There are also hundreds of billions of Galaxies out there that we know about each of them containing hundreds of billions of stars.

We must assume that even a small fraction of them will be found to have some form of life living on them because life itself can't be a random event either. Life is a form of beauty just as the stars and the galaxies are and this is why it's impossible for me to conceive that the Creator would have all of this design going on in his head without more things like us peppering the heavens on a regular basis.

There are as many galaxies out there as there are stars in our own galaxy. And we can only see a small percentage of them.

This is one square inch of the night sky where the Hubble Telescope took a time lapse photo of this tiny region in the night sky. They found that everywhere they do this, there are billions upon billions of galaxies each with billions of stars inside.

BUT - there's more. Recently we've found strands of energy that connect all of the galaxies in the universe so that it looks like the following.

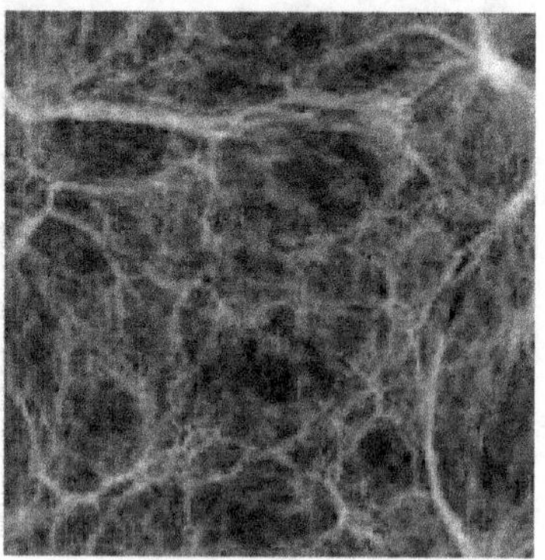

The human brain looks like the following. Can you tell the difference?

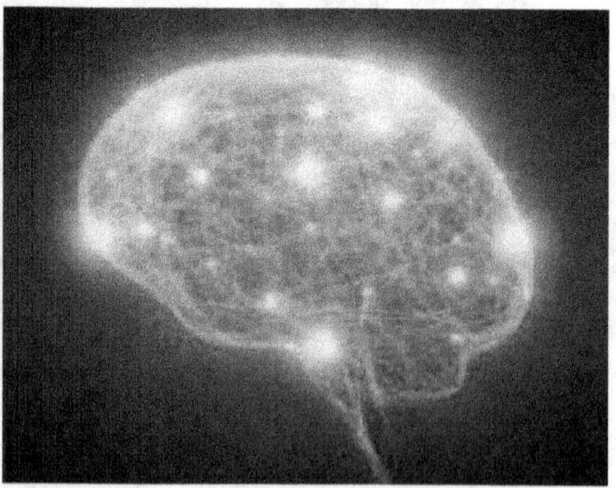

The God Particle Field creates every single atom that is

comprising these huge structures in the heavens that look and maybe even behave like the same mechanism in our own skulls - a Cosmic Consciousness.

You have to have a huge imagination to see in your mind what I am now describing to you. And for you to do this, miraculously enough, your imagination actually has to grow into an area much larger than the universe itself in order to contain this concept.

There is no number large enough to measure the number of sub-atomic particles and atoms in the universe. The number would have to be so close to infinity, you may as well use the concept of infinity. The same thing can be said about the numbers of brain cells inside your skull.

So, the way that I express this concept is a number that I created to paint the picture for everyone of just how powerful this concept of the God Particle Field and how busy it must be to keep an infinite number of particles in place so that the universe can look and feel and act in the uniform and trusted way that it does for billions of years now and counting.

The number is Zero to the power of Infinity. It looks like this:

It is a new number that I created to help us all understand how to unite the force of Gravity with all the cousin forces that are still perceived to live in the Standard Model all alone without the partnership of the greatest force that we know and use every day to remain on this planet - Gravity. It has to do with the concept of the size of our imaginations. In physical measurements, our brains are only a few centimeters in diameter and yet, the way we use our brains, our imaginings can contain the entire universe. Think about that for a minute.

It simply means that a Zero can be the tiniest amount of Space/Time and it can take up the largest amount of Space/Time in the universe. We only need to state where we are in our imaginations and the Zero Distance between all things becomes clearer.

The best way to explain this new number is to think about the force of Gravity. You've got all the information you need to fully understand the most basic construction of our universe. It's all in the Standard Model chart that I am putting here again for you to look at once again.

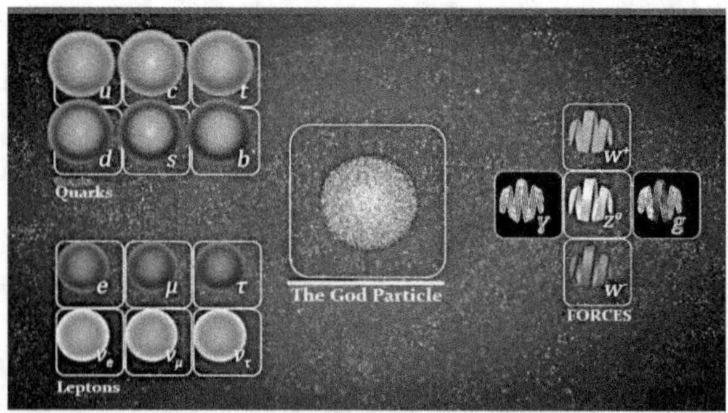

The God Particle is placed in the middle of the Stan-

dard Model of Sub-atomic physics because it is the force that gives everything else in the universe - existence.

This time when you look at this chart that summarizes all the fundamental building blocks of the universe and all of the mortar that God uses to keep them in place - notice that the other major force of GRAVITY is missing.

Scientists don't place Gravity here on this chart because they don't have any evidence so far of a particle that would carry the force of Gravity that we are calling 'Gravitons'. But, I believe that gravitons may never be discovered as particles simply because the force they carry is so weak, yet stretches out so far into the Cosmos, keeping all of the largest stuff in place, the planets and the stars and the galaxies that the particle that carries this major force will remain hidden from discovery perhaps forever.

Another way to put it is that gravitons are not like the other particles because they have to carry the force of gravity to such extreme distances, billions and billions of light years in distance. Whereas, the sub-atomic particles that we have discovered like the Electrons, the Muons, the Gluons, etc. are only charged to cover distances far less than the thickness of a human hair. It is this vast distance that Gravity must cover that we are forced to use this concept of infinity and zero all in one number.

This makes me think about Distance itself in a unique way. Distance doesn't really exist in the sense that we know it as living beings who have to cover distances every day in our scurrying about this tiny little piece of rock in the firmament of Space/Time. As living creatures we interact with the property of 'Distance' as it is an objective fact. But, for the particles that live in the sub-atomic world, there is no real distance, only probability of points in space which they will inhabit at different times.

Think about how different life would be if we were able to live like sub-atomic particles. Instead of getting into our cars or trucks or buses or trains to get to work every day, we would simply think about the probability of being at work when we want to and then simply induce ourselves into our workplace and 'Bingo' we would simply appear at work or at play or at shopping. There would be no concept of 'Distance' if we were able to live like sub-atomic particles. And hold on to your hats, folks because some sub-atomic particles live just like that - with no regard to distance at all. (Don't worry this is coming in the next chapter.)

Because of the force of Gravity, however, we are not able to live like sub-atomic particles because at our level in the universe Gravity is more influential than the other three forces because the other forces don't interact with this amount of mass. We're simply too bulky to be impacted by the other three forces in the universe that are designed to keep the atoms together and therefore can only work in extremely small distances. Remember, I told you that the forces in Nature that we know about are all whacky and counter-intuitive.

And, this is where my new mathematical concept comes in

handy. There is really zero distance in the universe. The illusion of distance is upheld by the force of Gravity. Gravity is actually the force that stretches all matter out like super-putty so that living things can live and breathe and interact with other places in the 'Super-Putty Universe'.

For this to happen, the number that can handle the smallest distances in the universe, so small that they are merely points of probability in Space/Time and the other expression of distance that goes out past anything that we can see with our most advanced telescopes and observatories, would have to be infinite in smallness and bigness.

This is Zero to the power of (i) or infinity.

$$0^{(i)}$$

This number was suggested to me when I was studying a math problem and something I had never known before popped up in the calculations and that is that any number to the power of Zero is ONE.

Take the smallest number in the world - even minus numbers and multiply it to the power of Zero and it is assumed to be ONE. Take the largest number in the world, multiply it to the power of Zero - and it is also assumed to be the number '1'.

It looks like this:

$$-.34171948374857362859995734267396352 78^{(0)} = 1$$

And/or it can look like . . .

$$386,496,633,794,622,692,496,339,194^{(0)} = 1$$

5 to the power of Zero = 1

5,000 to the power of Zero = 1

All numbers, large or small, positive or negative when raised to the power of Zero = 1

There is really very little difference between zero and one. Another way to put it is that Zero can always lead you back to unity.

I recently had an epiphany and have to express it here. Any number, including what we call 'Pi' with the symbol ,,

$$\pi$$

We should all know the famous mathematical formula for deriving the area inside a circle -

$$A = \pi r^2$$

Area equals Pi x the radius of the circle squared.

And the value of Pi as we can all remember is

3.14159265359 …

and it can be followed by an infinite number of numbers to the right of the decimal point because Pi is the ratio of the Circumference of any circle to its Diameter.

This means this number applies for any circle (And is also part of the volume measurements of a sphere) in the universe.

No matter how small, no matter how large any circle or sphere may be in the universe this number, this ratio of its circumference to its diameter is always Pi or …

π

Or ….

3.14159265359

Now, my epiphany came in the form of putting Pi into action in my previous calculations that any number raised to the power of ZERO = 1

So, that also means that Pi to the power of Zero = 1

This also means that any circle of any size raised to the power of Infinity can be any number because of this incredible ratio of Pi that is always the same, no matter where you go in the universe.

In other words, the circumference of the universe in relation to its diameter is still **3.14159265359** (to infinite numbers) because this is just one of the immutable mathematical laws we have to live by that was created at the first moment of Creation.

AND here is where you must take a Quantum Leap in your imagination. Zero is a circle or in 3 dimensions, a sphere. So, the circle of the entire universe whatever that may be if raised to the power of infinity is also '1' or one.

Any number whether positive or negative, the largest number you can conceive or the smallest number you can conceive raised to the power of Zero all equal the same number of '1'.

AGAIN - if any number to the power of Zero always equals ONE - then Zero to the power of Infinity equals ANY NUMBER. Expressed mathematically this way.

$$O^{(i)} = U$$

Zero to the power of Infinity EQUALS a Universal Set of Numbers - or ANY NUMBER no matter how large or how small the number.

Another way of reading this formula is that Zero to the power of Infinity EQUALS the Universe. 'U' can stand for any number in the universe, or the universe for short. They are interchangeable. This may have serious implications, by the way, for the 'String' theorists, which we'll discover later.

All of God's creative talents exist in Zero Space and Time because in this space and Time he or she can be or think or use any number, no matter how big or how small and things get done instantly without any limitations due to distance. This allows for the incredible harmony we feel in our gut whenever we discover another of God's great truths. God maintains our universe perfectly balancing all forces and particles with a simple wave of his hand at the most granular level or at the largest macro-level all with the same brush strokes, and with no wasted effort.

This is how God creates and maintains the universe we inhabit. God can be and is any number all at the same time and this is the great Cosmic Symphony that we perceive to be true, because it is truer than true. And, it has been hiding in plain sight all since the beginning of Time.

It should be easy to remember my formula by thinking about it as an 'I.O.U.' Either that the universe owes it to you to be supportive of your goals and ideals. OR, the better interpretation is that you OWE it to the Universe to understand and fully appreciate your role in it. Either way is fine with me. I personally think about it as I owe it to the Universe to be the best version of a life form that I can be, because I can appreciate just how hard the universe has worked to create me and maintain my home planet in this wildly cold and warm but mostly majestic place.

And Zero - mathematically - is just an empty space in the way that numbers get bigger or smaller. When ancient Humans, we think it was the Sumerians discovered the ZERO, it was some early mathematician who realized that we needed a 'Place-Holder' and so the first Zero in human history is written like so:

<>

This was to imply a place where there would be no numbers and you start counting again from that place to the next one. As we count to 10. '0-1-2-3-4-5-6-7-8-9-1 with a second 0' and so forth.

Today, we now see that the place that Zero is holding is not just a tiny little place but the Infinity of Space and Time itself.

This formula is what I call the **Universal Constant** and mathematically it means that the entire universe can be as big or as small as you would like. I know this sounds more than a bit crazy, however, what I am telling you is now a known fact as we have recently measured with our instruments at the LIGO observatory the actual 'Warping' of the fabric of Space/Time. If the entire universal multi-dimensional fabric can be warped, it's easy to make the logical assumption that it can also change its dimensions in any manner required under the Laws of

Physics.

And if Einstein had used this formula instead of his own Cosmological Constant I believe he would have proved his Unification Theory in that Gravity would now be placed into the Standard Model of Quantum Physics as one of the Four Forces, albeit it would be negligible still in all calculations for 'mass'. My Universal Constant is similar to water, the Universal Solvent - anything, any mass, any number can be dissolved into it or exploded out of it.

Of course, there is no largest or smallest number in the world, these things are infinite in both largeness and smallness. Numbers can go to infinity and that means that zero can too. But, now that we know that zero can also go to infinity, it means that the four forces can work over the largest and smallest spaces in their own missions of upholding the structure of the universe for us because all distance in the universe can be considered to be really Zero to Infinity. It can be infinitely small distances or infinitely large distances, it all end up in the same place, and this is the so-called fabric of Space/Time that Einstein introduced us to by describing how one force - Gravity - shapes it like a sculptor making his pots.

Zero to the Power of Infinity equaling any number means that you can substitute any number for any place in the universe when you are calculating a place or a distance from one place o another. The Beatles got it right. Space/Time is a huge 'Rubber Soul'.

I believe that this new concept will take some time to sink

in for most of you. It took me nearly all of my life. It started in High School when I studied Euclidean Geometry and my books taught me how to calculate 'the shortest distance between two points' and how this was defined as a line between 'A' and 'B'. From there we are able to measure the sizes of everything in the universe or even the area or the volume of things. And if you merely take these lessons in Euclidean Geometry on their face value, you will do well in your calculations. But, later on in my life I started to think about the distance contained inside the 'Two Points'.

In other words, your little dot on the paper that is one of your points in your geometry, have size or distance. They are so small that you can say that these distances are negligible and don't have to be included in measuring say the distance between New York and Los Angeles, or the size of your windows. However, the true distance to its greatest accuracy cannot be calculated because no matter how small you make the beginning point and the ending point of your two spots on the map, they will always have a size or distance. It's the size of your pen or your ink spot on the page. You can get it down to the size of a sub-atomic particle and now you only have a point in Space/Time that is a probability.

Therefore, all distances between all things in the universe are only probable distances. Nothing is absolutely tied to one another. And, this is how and why my new number . . .

One more bit of information may make this concept of zero to the power of infinity come into focus for you. I alluded to the way that the God Particle creates another particle in the universe a few pages back, but I left off the best part of how the God Particle Field makes matter come out of the zero distance of Space/Time.

I think you'll find it more than just interesting that the thinking machines that we have come to rely on in every aspect of our lives today, the computers, are designed to figure what to do based on zeros and ones. As we all know, computers cannot count past the number '1'. In computer thinking they know about 'On' a Zero and 'Off' a '1'. This is known as the 'Binary Code' in which all computers are programmed. That's all they know about and they help us make it about in our world by simply counting how many things are on and how many things are off.

When you look at the sub-atomic world, every particle that we know is either 'On' or 'Off'. It's either alive or dead, present or not present within the atom. They turn 'On' and 'Off' based on their interactions with the God Particle Field.

How does the God Particle field turn all of these parts of the atom 'on' and 'off', alive and non-alive in our world?

And this is the most interesting question that we can ask ourselves at this time.

The Answer:

The God Particle merely gets things to resonate with it. The resonating or vibrations of a particle's wave energy is how all the basic building blocks of the universe come into being and how they stay 'ON' or 'ALIVE' in the physical world. They are simply turned 'On' by the God Particle Field or turned 'Off' when no longer needed. Their existence in this universe is binary, the same as all living things.

When they are turned on, all sub-atomic particles continue to hum in a musical pattern that remains perfectly vibrating to the note that they are supposed to sing. Some particles might be singing in the key of 'C' and others are singing in the key of 'G'. Others still are vibrating to the key of 'E' or 'F' and some are even playing along as the Flats and Sharps of the musical notation. The music notation of course is much more complex than our simple system of the octave or eight notes. The universe is playing along in an orchestral performance that is unmatched anywhere. It is purely in harmony with itself and with every living or non-living thing that the music creates.

And, the more music that is playing in any given region of Space/Time, the more things are going to happen. Galaxies will come into being where the music is strongest and empty Space/Time with zero distance will remain in place where there is very little music happening. It's no more complicated than that.

48

Where the master composer has called for an Oboe, an Oboe will sound in perfect key and it will be part of the music. Where there is a need for saxophones, saxophones will lean in. Where the master composer has called for a violin or a sea of violins, the violins will resonate with the written music as it is told by the score and is placed on the music stands in front of them. And the concert that was started thirteen or fourteen billion years ago is ongoing and is building in its crescendo and in its passion for life. Can you hear it?

To the extent that some or any improvisation is allowed by any of the players in this concert, this will be discussed in the next chapter.

<u>Chapter Two</u>

- Improv

In the last chapter, we learned how the universe works in great detail. Most of us don't even need to know about how the universe works in this amount of granularity because all of that detail seems so complex and mind-bending to many of us. But, today, we have so much information on the tiniest building blocks of the universe, it makes for extremely interesting conversation at cocktail parties, at least if you have friends who are interesting and want as much knowledge about their environment as they can before they depart this train station for the next point in their itinerary.

And the best part is that it really is ALL very simple. You just absorb what you can absorb from a discussion like this. You will get what your brain is capable of understanding at this time. Then, later, perhaps by reading a book with competing ideas or series of articles or TV documentaries, it will all snap together in one neat little package as my perception of the universe has done over the course of my lifetime and with the help of all of the above, plus friends who would not give up on me.

In the back of my mind, I also have the help of one Albert Einstein because I learned decades ago that he famously said,

'The only incomprehensible thing about the universe is that it is comprehensible.'

In other words, it was amazing to Einstein just how simple the universe really is and he backed it up by explaining one of the most incredible discoveries ever made in one simple equation:

$$E=mc2$$

We all know today that this simple little grouping of letters stands for the fact that all of the Energy in the universe, a little bit of it or huge amounts of it the 'E' in the equation is exactly equal to the mass in question, again whether it's a tiny little electron mass or the mass of the brain in side your head 'm' multiplied by 'c' which stands for the SPEED OF LIGHT and the '2' means SQUARED or times itself.

The speed of light, Einstein had already discovered is a constant, which means that everywhere in the universe it travels at the same speed, which most of us know today is 186,000 miles per second. That's screaming hot fast - right?

We also know that the speed of light is the ONLY CONSTANT in the universe and that everything else is relative to

this constant bit of mathematical wizardry that has been measured in the laboratory - so we know it's true.

But Einstein's formula, which led to the development of
atomic bombs and this is how we know how correct it is, is also
telling us something even more interesting. That matter and
energy in our universe are really the same thing, they're just related by one amazing part of our universe - light-speed, or
maybe even light itself. And if it's possible for matter to turn
into energy with explosive results as in an atomic bomb, imagine for a minute that the universe can transform itself with im-
plosive results by squeezing all of infinity into our tiny little
brains. It just takes 'Time'.

(- NEXT POINT IS IMPORTANT. Please read this next
part slowly until it sinks in.)

So, Einstein has proven really that energy and matter are
actually the same thing. The only thing that makes them different from our perspective of the universe is the speed of light,
which is constant everywhere and a very big number. However, from God's perspective, the speed of light is extremely
slow. Therefore, this relationship between Energy and Mass in
our universe is really just a simple and uniform way for God to
slow things down everywhere just enough to impact our world.

And, if the speed of light is the differentiating factor between all Energies in the universe and all the material things
that we encounter in the 'Real World', then Zero to the exponent of 'Infinity' is the only differentiating factor between all
things in the universe both great and small. This may also be a

proof for the String Theory as well, because I'm starting to see the 'Strings' that String Theorists are suggesting might be the tiniest building blocks in the universe, as tiny circles or Zeros, not real strings because when you think about it, if these things were actual stringed structures, you'd have to ask 'What are the 'Strings' made out of?'

A true Zero, on the other hand, holds nothingness inside it and nothing more, therefore, we never have to dig down any further.

(End of section, I need you to absorb before going any further.)

Now - let's go back to the beginning. We can become more familiar with this concept when we look up at the night sky. All of the light that we see emanates from the stars and planets and galaxies in our region of Space/Time. But, the distances are so vast between stars that the time it takes for the light from even the nearest of stars in our own neighborhood group of stars at least 5 years. The light from the farthest stars in our own galaxy will take thousands of years to reach us. This means the closest stars are 5 or more light years away from our sun. Or it takes 5 years for the light of our closest neighboring star to reach us traveling at the incredible speed of 186,000 miles per second and indeed this known speed of light is how the Astronomers and Astrophysicists are able to calculate the size of our universe and many other properties of this place.

And all of this means that when we look up at the night

sky, we are actually looking back in Time.

So, the earliest we can see from the universe around us is what happened five years ago, unless we're talking about our own solar system's planets which would be from a few minutes ago in the case of our own moon to several hours ago in the case of Mars.

In terms of the furthest stars and galaxies from us, we have built instruments like the Hubble Telescope that can gaze back billions of years in the past by revealing to us the light that has originated from galaxies that has taken billions of years to reach us.

Amazing, but true. When we look at the largest aspect ratio of our universe, we have to look back in time. We have no idea what's happening in the universe right now. We can only make assumptions based on information from the Past. Conversely, when we look at the smallest aspect ratio of our universe, time no longer exists, as we know it. In the sub-atomic world events are moving so fast that time is almost meaningless. There is only the probability that something will be where we think it could be in the future. We have no idea what's going on in the smallest regions of Space/Time any more than we know what is going on in the largest aspects. Luckily, everything that we see in the Past by looking outward and everything that we learn by looking inward seems to be persisting at the moment, and so we assume that everything is the same here in the Present as it was in the Past and as it will be in the Future.

The wave action of an Electron, for example is a number

with 15 zeros after it or quadrillions of waves per second, depending on where you find them. Well, that's not a measure of Time in our own layer of the universe that is meaningful for our brains to consider in our perceptions of reality. Try to imagine the ocean waves coming at you at this rate of speed. You can't. For laboratory work, this information is essential, but for us as living creatures trying to see the big picture it's really useless information in terms of everyday life matters.

Now, when an electron jumps from one orbital to a higher orbit around the nucleus of its atom, it must ABSORB a photon of light energy. And, when an electron falls to a lower orbit around the nucleus of the atom, it must RELEASE a photon of light. So, the universe we inhabit uses light to send us information about the Time it's been around and it also gives us light waves to measure the 'mass' of an atom. But remember, the 'mass' of anything is really just a bunch of energy related by the speed of light squared. It will take a very long amount of Time for a light photon to travel the distance of 186,000 miles X 186,000 or a total of approximately 34 and a half BILLION MILES.

Another way of stating this is that there is 34 and a half billion miles between one mass in the universe and the next highest mass. This would be a very difficult calculation to be making every time we want to weigh a pound of bread. However that's where Zero (I) comes in handy once again. The universe will stretch or shrink to fit anything we need. Not a shabby trick if you can do it.

Time, in other words is never really what we think it is.

There is no present moment because the very second you can put your finger on the present thought or action, it has gone into the past. Our brains do not work fast enough to capture the true essence of time which is ever moving and changing from the Past to the Future.

Oddly enough, however, our brains are quick enough and intelligent enough that we can design instruments that help us look into and examine carefully all the elements of the universe in great detail to give us a snapshot of where we are as we are sandwiched quite neatly between the two aspects of time that we can see, feel, taste and touch with our minds.

The main train of logic here is to show you how easy it is to get lost in the weeds. We want to stay out of the weeds, out of the detail because it can take up so much brain power that the larger lessons are missed.

We can and should recall often the greatest adage of all - "Too bad - he is just too close to the forest to see the trees."

In this book, we're acutely aware of the forest and how it works to support the trees and other life within it, but we want to stand apart and look at the overall beauty of the trees and how they relate to one another and even to ourselves. If we lived totally within the forest, for another analogy, we may never realize that it is the trees themselves, and the other plant life, that scrubs the air of the poison gases we living things pour out into the air through our breathing and digesting activities and returns to us all of the essential oxygen that we need to breathe in, to feed every cell in our body and thereby remain alive. This is known to us as the great cycle of Life.

In this same way, the great cycle of Space and Time coming out of nowhere and creating so much of this nothingness that we can depend on some of it to feed us, clothe us, house us, educate us, support us in every way we may require. Energy making more of itself to give us the energy to do everything we need to understand it all. This is what Einstein meant about how amazing it is that it's so simple.

$E=mc2$

because ...

$$0^{(i)} = U$$

0(i)=U   My Formula 'Zero' to the exponent of Infinity equals any Number. This means that Zero or nothingness can be any size – no matter how big or how small. When I created this concept, I wondered if this is how God has organized the universe so that he or she can be ubiquitous and give us everything we need to survive at any time. Think of it as the biggest I.O.U. of your life - exactly what you need coming to you from the universe whenever you need it. Can you think of a better way for God to organize the universe so that he or she can be everywhere at all times in order to give us everything we need to survive while God creates a debt that we must pay back at some point? I cannot.

If you understood the sketch of the universe that I made in Chapter One, it may seem to you that everything is pre-determined and that the action of the God Particle tells everything where and when to pop up out of the nothingness and begin playing a role. It begs the question:

'So, why bother to even try to be a 'good' person with individual goals and aspirations if everything in my life is going to be set in motion and continuously monitored by some great hidden force of Physics that I have no control over?'

GOOD QUESTION:

And so, I'm here to tell you that the following is where Improvisation comes into play. The Creator of Heaven and Earth knows that his or her majestic symphony of the universe would be static and boring unless it allowed for some of the musicians to improvise from time to time in a collaborative effort to make the symphony constantly entertain us and keep us in our seats so to speak, happily singing along. When we recognize just how beautiful the Cosmic Symphony really is, this is where our inbred talents come to the fore.

Any single person, even the animals and even the insects, all living things on this planet can join in the Cosmic Symphony by simply opening one's eyes and ears, becoming familiar with the beat, going along with the flow of the melody, predicting the phraseology of the score and what might come next - AND THEN THE BEST THING IN LIFE CAN HAPPEN - you can feel it. It's in your heart and soul. It's a nod you get from the Maestro when he hears your voice, your instrument, your soul resonating to the music and then pouring forth with your own musical notes.

After a few clumsy attempts to add to the symphonic sights and sounds, you gradually get more and more in the way of encouragement and musical instruction from the Maestro, our great symphony orchestra conductor. And, your contribution grows toward a crescendo in your own life as well as in the life of others, if you're lucky enough to have that much to offer.

I can assure you that this recognition of the music coming from the Cosmos and impacting my life is what has led me to this particular work of mine, this book, several other books, as humble as all of my life's work may seem, as much ridicule as I may get, with all of the criticism and push-back that I may receive, this is my contribution to the Cosmic Symphony.

Now, some of you are wanting more detail in the Physics of what I have just described as the musical improvisation which is your life. So far, I have given you a great deal of the scientific basis of every one of my theories, which I hope you have appreciated so far. I've tried not to get too detailed so that I don't lose any of you, and so if I have lost you, I apologize, but in that case, I would request that you go back to the beginning of this book and read it again to this point because this is where it gets really, no really interesting.

But, I will give you the Scientific basis of this wonderful uniqueness that is YOU and I can do that by having you recall that I said that the position of the sub-atomic particles is never truly in one place. We can't know where everything is in the sub-atomic world. The particles, the electrons, the protons, the neutrons, the quarks, even the muons and gluons can only be held to a probable position in Space and in a probable position in Time.

This probability of existence for all matter that the Master Coder of the universe has included in his design was made part of the Science of Physics that we are here studying today to some degree so that WE WOULD HAVE our Free Will, the ability to play along with the concert or go off on our own direction. Some of us may even thumb our nose at the Maestro himself, daring to try to be as great or greater than the master conductor himself or herself, if there be a gender to the Maestro. This is futile and dangerous and when we go off in this direction, we may lose our soul. Even so, many will perish or become lost in the attempt because of their over-developed egos.

Now you have to know this too. God (Please Note: We can not know God directly - From these footprints we can only know that God exists, and how he or she operates, not much in the way of a physical description) could just as easily have created the universe with no concern for free will and we humans along with a lot of other creatures out there could have evolved to just go along with the program and never express any creativity, no musical improvisation.

AND THAT'S HOW I KNOW that God is filled with LOVE for all life forms everywhere, yet favors no individual, nor does he or she conspire against any of one of us. Only a loving Creator of the universe of this grand and sweeping scale would design into His Plan the laws of Probability where all matter can come and go, appear and disappear at their own desire to 'Resonate' along with the music and take part in it. We living things are no greater than any of the smallest non-living things. Only a Love of this size and perpetuity would plant enough of it so that any of us compiled from this matter could go our own way, making our lives whatever we want them to be. And this is a good thing, is it not? Would you want to live any other way?

Would you prefer to be a robot?

These are NOT the Droids you are looking for.

Remember, the way that all other matter in the universe is created is through the act of a 'RESONATING'. Somewhere in the sub-atomic world an Electron, Proton, Neutron, a Quark, any energy that forms up the sub-atomic world and which builds up to our perceived reality simply begins to 'Resonate' with the God Particle Field. I can't state this enough. How can something so easy be so ubiquitous and so automatic that it has spread itself all over the universe and is continuing to spread outwards to infinity to this day?

We may never know the answer to that question but this is the one question that we need to know the answer in the Science of Physics and is so far unanswered.

So, let's take this problem apart and analyze it step by step. First what does 'Resonating' mean?

Whenever you listen to music and it makes your eyes water and you feel a warming in your heart, you are 'Resonating' to that pattern of sound waves that we know as 'music'.

In Physics, we can see the power of 'Resonating things' by looking at the tuning fork.

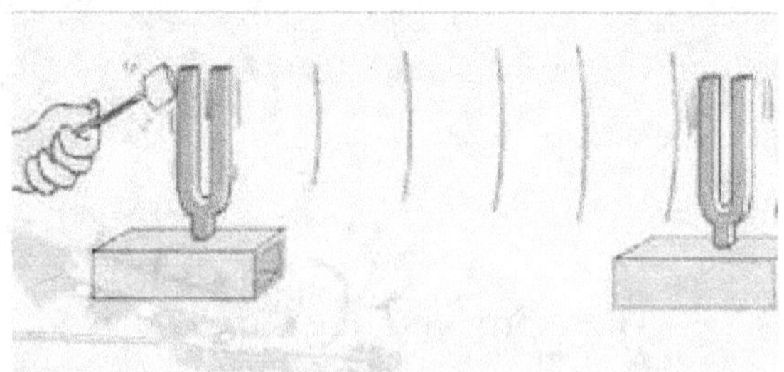

When you strike one tuning fork, another one even far away will start playing the same note as the one that you struck as it 'Resonates' with the struck tuning fork.

This is of course what happens whenever we create music:

Music is simply the Resonation of the Instrument and then the Resonation with the listener.

What makes a person resonate with the music? Who can say for sure? I would postulate that it has to do with the layer of existence that is embedded in every living thing and in some cases non-living things that exists in the universe and this is known as 'Consciousness' or an awareness of who we are, where we are in the universe and where we're going - as the fullest expression of awareness.

Many people I have known have only a superficial under-standing of this property of all life forms. They are still aware of themselves in their skins because this default amount of awareness is conditioned within us as children when parents give us our names and start to train us as to their expectations.

After childhood awareness, it's really up to us to advance our level of awareness to any heights that we believe to be our birthright. And, the way in which we all raise our awareness is through our 'Resonating' with ideas. If something we read or hear sounds like it's true, we adopt it and include it in our own regimen of ideas that we can defend and call our own. In this way, we are resonating with Truth. If something does not appear or sound like the truth to us, we reject it, but even in rejecting certain ideas, we define more of the edges of what is our own ego. We might say or think - "We can see right through them." or "That person is an idiot."

Even in rejecting ideas or proposals put forth by others, we are resonating with those rejected ideas, but in the negative sense and our awareness of ourselves, our ego, keeps score. Add up all the loves and hates, all the acceptance and rejection of varying truths and this is who we are at the end of the day. When we lay our heads on the pillow our thoughts about the physical world around us fade and the person who resonates with the truth about his or her life, arises in our Consciousness and gives forth to our nightly dreams.

And this is why it's so important to dream and dream big. With every dream at night or even more so in your day-dreams, you are helping to define the real 'You' who can then resonate with the real music that is aimed your way. Yes, everyone hears different music out of the same Cosmic Symphony. The music that is the universe is everywhere at once, but it's also fine-tuned to every individual living soul, so that it actually sounds different to everyone.

This is what we might call our 'Taste' in the Cosmic Symphony. Remember, the sub-atomic particles can have 'Flavors' and that means that the mix of all these flavors in our own particular realm of the universe is our own 'Mix' and we perceive our own music according to our 'Taste' in music.

With this current understanding of the Science of all the Physics that matters, we can now look at the universe in terms of our own personal listening booths.

When I was a kid, there were things known as 'Record Stores' and inside of these were private listening booths. We would go through the records in the bins in the stores, choose a few vinyl records and then take them into the listening booths that were sound-proofed so that we could turn up the music that we loved very loud and we could 'rock out' to the Beatles or the Rockettes, the Everly Brothers, Buddy Holly or Chuck Berry.

The universe gives each and every one of us a 'listening booth' a private version of the music that seems to be written for us, where we can enjoy the sounds of music in the volume and the genre that we like and resonate to. Over time, our listening booth gets more and more specific in the range of musical experiences, because the universe learns what kind of music we like by the number of times and the manner in which we resonate to it. Therefore, our Consciousness is a two-way street.

We get information about ourselves from the universe and we feed back to the universe what we have discovered in our listening booths and how much we know and love different pieces of the overall symphony. The growth of our Consciousness then, is guided by the nature of our listening booths, in how much we use them, and what we are listening to and resonating with. For everyone, it's different. And, this is where our own Free Will, our own improvisation of our own music - our contribution to the sounds all around us - takes place.

These old-fashioned listening booths from my era have been replaced by a personal listening device that comes out of the Internet and can be made into a personal playlist of songs that we appreciate again as unique individuals with a completely unique set of tastes.

Thus the 'Flavors' of the sub-atomic world are passed on to us in the quantities and the quality of music that is custommade for each of us. Then, we are allowed to react according to our own individual 'resonating' thoughts and patterns that will give a constant flow of information back to the sub-atomic world that we are all composed of. Upon receiving this information about our own tastes and lack of same, the universe will adjust and give us different music than before.

So, the way in which we listen to the cues that hit us every day from 'out of nowhere' - or so it seems - the way that we react to them, is our improvisation of the music that is part of the universal symphony as it's being performed 'In Toto' by the Maestro, the conductor and the magnificent writer of all the music that makes up everything else that we see and feel.

Who we are as individuals is composed of the music that we resonate with and nothing more because as we resonate, we feed back and the stream of music is adopted to fit the new and constantly changing versions of ourselves that is called our 'Growth'. And this is based on our free will to react in any way we see fit.

This is how free-will is programmed into the universe. Make no mistake, this is how we know we are unique and how we know we are not automatons or robots. We can choose to listen to whatever music we want to in life and that responsibility for our taste in music is completely ours. What is the variable here is that the system is open to any ideas that you may want to propagate throughout the heavens. It's all going to be part of the Symphony, in good or in bad taste. It doesn't really matter and this is why it is important that we, as individuals never judge the life choices that others make. Nor should others judge our choices.

The system is designed to give us all the freedom to choose whatever it is we're choosing in the friends we keep or the actions we take. If we judge some of these as inappropriate in any way, it's only our ego speaking to us, that has been over-worked, over-emphasized and it has fooled us into thinking the ego is us. It is not. No one individual is superior to any other. We just think so at times from an egotistical point of view. Banish the ego from your own Consciousness or at least recognize when it appears in your mind and you will free yourself to resonate more often and with far more enjoyment because this is your true self.

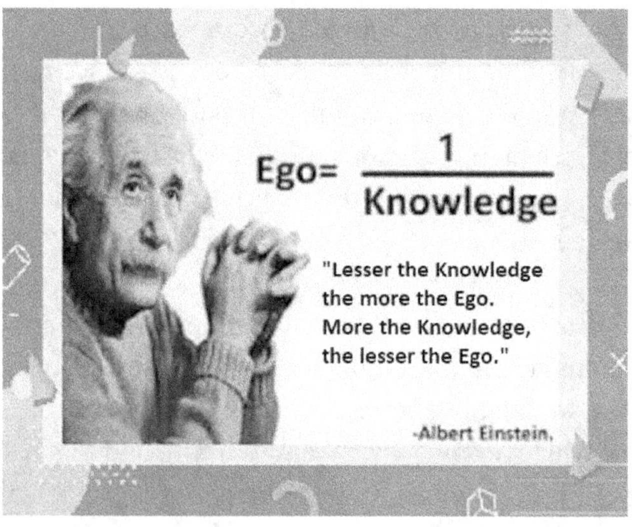

So, now you are prepared to go out into the world and know how the universe is working all around you. You are inside a kind of cosmic 'Listening Booth' where you can listen to music that is of your own taste, while flavors of the universe are playing out to a different 'Taste' in and among the people of your friends or family.

And if you want to become a truer friend, or a more loving

lover, every once in a while drop in, and tune in to listen in to the other person's music. You'll be amazed, enlightened and more 'In Tune' with that other person, or all the other people in your life.

<u>Chapter Three</u>

- Awareness

Throughout this book, I have used the word, 'Consciousness' in many different ways. There is a Cosmic Consciousness that I talk about as if it is as close to a definition of 'God' as I can find. There is also the Consciousness inside your head and mine. And over the course of my lifetime, I've struggled with how to connect the dots, so to speak. Is there a connection in the universe between our own level of Consciousness and the Consciousness of a 'God', a Creator of all things?

In the first chapter, I defined exactly where God exists and proved it mathematically. In this chapter, we're going to consider how and why the Maestro uses this type of 'Hiding In Plain Sight' formula for our success and survival. Remember, like any music, resonating with this information is the key to your total enjoyment of the next several minutes of your life, or any other segmentation of it.

When I stumbled upon a set of experiments that have taken place in the last fifty years of Science, I believed that I had found my answer to this age-old question and to this day, after thinking about it over and over for multiple books like this, I am now convinced that I was right to connect the dots in this way.

It has to do with a discovery of a proof of an Einstein theory made by a Physicist Jonathan Stewart Bell. When I first read about 'Entangled Electrons', it was very confusing to me and I had to re-read the arguments over and over. Over the years, it finally sank in to my brain of the importance of this - probably greatest discovery in Science of all time.

We're mostly all familiar with Einstein's famous theory of relativity in which he claims that nothing can travel faster than the speed of light, and in most things, this law of Physics appears to be true, although not in the movies. Einstein theorized that since nothing can travel faster than light, the CAUSE of some event couldn't either, because this would be backwards in our intuition in believing that the Cause of something comes first and then the 'Effect' happens, and not vice-versa.

And this is what gives us our sense of the universe where we can watch one thing bring about another event, every time and it's never the reverse, where a later event causes the earlier one. Star light for example, always comes after a star is born. The light never happens to reach our planet where we observe the starlight and then, some days or years later the star is born. Things just don't seem to happen that way in our universe.

When others argued that it would be possible for some things to happen first that were caused by an earlier event, Einstein called this 'Spooky action at a distance' and most people forgot about it. Until one day, when it was observed that two 'Entangled Electrons' could change their relationship over huge distances and there would be no time lost in the causation. One Entangled Electron can change it's spin when its partner's spin is changed somewhere else, but the second Electron takes zero time to do so as if there was some message between the two electrons that traveled across space faster than the speed of light, with no delays.

This has been tested so far to the distance of about 10,000 miles when one electron is on the Earth and its partner is launched into orbit. When the first electron's spin is reversed in a laboratory on Earth, it's partner electron, 10,000 miles away in space will change its spin in the opposite direction IN-STANTLY. If information were traveling at the speed of light between these two electrons, this reaction should take place several seconds later. It doesn't. Therefore, when Einstein saw this evidence first produced by John Bell, he was forced to change his mind about the action of all cause and effect in the universe.

So, after learning that even Albert Einstein could be wrong once in a while, I had the temerity to think about this discovery in a more general way.

I thought to myself, 'Well, if one of these electrons changes spin when we change its partner's spin somewhere and these two little sub-atomic geniuses are not connected by any mechanical means - how does the partner know about the state or condition of his her buddy?'

And, how does it know, instantly, without any time to think about it? Even more wonderfully, 'Why does it change its own spin in reaction to its buddy?'

The only answer that I have found is that electrons can be aware of another's existence. This awareness of the other also means that it is aware of itself first. An awareness of the self is the only clean definition of 'Consciousness'.

I hear a little voice inside you saying 'Oh boy, now he's going to say that electrons have Consciousness'. And, not to disappoint you, but that's what I have stated in an earlier work - 'The 4 States of Consciousness'. The first state of Consciousness, it appears, is the state of awareness that electrons have for one another, now proven in the laboratory, but only when they have become 'Entangled'. Entanglement means that they have been shoved so closely together that some property has rubbed off between them. I call it a marriage. They had a ceremony. Someone pronounced them 'Entangled' and they remain faithful to each other for the rest of their lives like swans.

Don't ask me if they have children. I like to think so, but there is no evidence of that blessed event in the universe - yet.

If and when they discover Electron babies, I would like to be their Uncle!

If you want to take on the challenge, try to think of any conceivable way that one sub-atomic particle could change its direction in life, at the exact same moment that its partner changed its direction. There is no way that one can 'KNOW' the condition of the other without awareness of the other. There has to be a knowledge of this somewhere or there would be no observed changes in either electron. If you can think of any other mechanism that would make sense given our understanding of how the universe works - please let the world know. You could be the next Einstein.

Now, at this point, it's important for all my readers to understand the difference between 'Consciousness' and 'Intelligence'. Consciousness for the purpose of this book at least is the 'Recognition of one's existence and one's position in the universe.' When we are little babies, very quickly, we gain an awareness of ourselves because we're born hungry and so we go looking for Momma's breast and if we're lucky little suckers, we are breast fed and it's because we know that we exist. We don't know much more about ourselves, but we know enough to seek out Momma's breast to get the earliest form of nutrition that we can enjoy and grow with. (I hate to shock some of you, but we are mammals first and above all.)

As our brain develops normally, because Mom is taking such good care of us, we slowly gain more and more intelligence. We learn how to eat other foods our parents give us. We even start learning how to communicate with our family after a very few months on the planet. Over time, we gain enough intelligence to read and understand concepts of Physics like the ones you are reading about here, a very advanced pile of knowledge that has taken thousands of years to produce.

But, and this is a BIG BUT - no matter how intelligent you become you are never more aware of yourself than you were at birth. There can only be one of you and one size fits all. There is only one series of events that created you. You can only become aware of yourself to that degree, that you exist. Even if a million or a billion people are aware of you and spread your fame far and wide, you can only be conscious of yourself as proof you exist, but that is all.

Your intelligence can improve over time as you grow and mature, your brain gets bigger in your skull and you pack in more and more Science from your teachers and your reading. But, you're awareness of yourself - your Base Consciousness never increases. It never improves. There is nothing more to understand in the understanding that you exist.

It's binary. You either exist or you do not exist and therefore your awareness of your existence is on or off. There is no in-between form of awareness of self in relation to the universe. There is no in-between levels of Consciousness. When you exist you are aware of existence and when you no longer exist, you no longer have Consciousness of self as you had existed.

BUT - and this is an even BIGGER BUT - when you yourself in your present form pass away from this existence

composed of a universe of Electrons, you will not become less
Conscious. Your Consciousness after death will instantly transform to the Consciousness of the Electrons that made you walk
and talk and roam around in the world. And remember the Formula of mine. When you die, your energy goes to Zero and
Zero is the place where God hangs out.

How can I say that with such certainty? Because for me -
Consciousness is the one thing in the universe that is immutable.

If electrons can be proven to be Conscious, then probably
everything is. We just haven't discovered everything in the
universe to that degree and when we do, it's going to be a long
but exciting road. I believe in this so strongly because when
you look at the greatest discoveries in Science there is over and
over again the evidence of an unseen hand that takes part in all
things.

Not everyone sees this hidden hand and so I need to explain it here. The greatest way to explain this concept is from
my earlier book - 'The 4 States of Consciousness'. You've already learned about the First State of Consciousness that we
know about and that is the awareness of one Electron for the
existence of itself and its partner.

But, what do Electrons do best? What is their greatest role
in our universe? What is their highest and best use? Electrons
are there to join with other sub-atomic particles, mainly the
Protons and Neutrons. And now we know these little things are
made from Quarks, at least three to a pack. And when the

Electron joins to these other little things, they form a union of energy that we call the 'Atom'.

Now, the Atom doesn't like to be floating around all alone either, so it combines with other atoms and the energy that it uses to seek out and combine with other atoms is known as the 'Rotational' energy, or the SPIN momentum, like a coiled SPRING with much stored energy inside but held back. But the rotational energy forces the Atom to rotate just as the Earth rotates around in Space/Time to create night and day every day, three hundred and sixty five rotations per year, and all of those comprising one single rotation or revolution around the Sun.

Since, this Rotational energy is given to the Atom by the Spin of the Electron, we have to conclude that the Consciousness of the Electron has been passed up to the Atom because nothing in the universe will want to give up Consciousness. Remember, it's immutable. It changes form, but never dies, just as matter simply turns into energy, when we die. Atoms, therefore, know about their existence in the universe and this is why they seek out other atoms to form Molecules. The most famous molecule being the one we need so sustain life on this planet - the molecule of water - we all learned as H2O which is scientific notation for TWO HYDROGEN ATOMS and ONE OXYGEN ATOM.

Water, a liquid that we enjoy flavored so many thousands of different ways, the stuff that is so delicious and life-giving is made out of two gases, Hydrogen and Oxygen that magically come together to make a liquid.

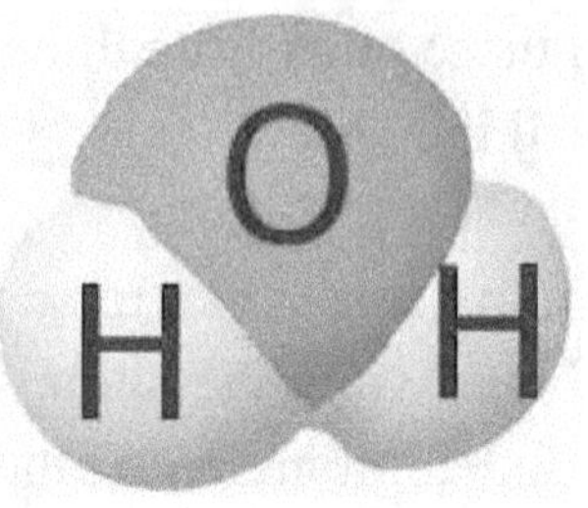

H2O (water)

The two Hydrogen and the One Oxygen Atom are each individually made from the 1st State of Consciousness found in the Electron, but is passed up to the Atom in the Rotational exchange of Consciousness because the Electron knows about its own existence but wants to play a bigger role in life. We aren't making this up. It's proven scientific fact today. The Rotational energy of the Atom induced by the Electron that makes it is now the container for Consciousness.

And aren't we lucky that this can happen in the universe, or else it would be a very dry and dark place indeed would it not? This to me is the first indication of the hidden hand in the mechanisms of the universe, what I have been describing to you is the dedicated work of the Maestro's, the conductor of the Master Symphony of all these tiniest of things, each and every one of them playing his or her part in the majestic symphony which we can see, taste, touch, smell and hear.

But wait! There's more - so much more! As we all know the little molecules of water and other ones like the Amino acids, the proteins - even our DNA are more complex than this simple molecule of water that make up most everything else.

So the little molecules combine to make bigger molecules and this is from what we know is the Vibrational form of energy that allows molecules to seek one another out and combine to create more and more complex molecules, like the ones that create our CELLS. All living things are made from these more complex chains of molecules, that are all classified as part of the study of 'Organic Chemistry'. These organic chemicals that make our lives work so flawlessly and seamlessly from the interaction of trillions upon trillions of little molecules, are the proteins - proteins that help us see, proteins that allow us to digest our food, proteins that allow us to lift heavy things, do technical work, to walk, to run, to play our sports.

So, the energy used to make us and all other living things is the Vibrational Energy that has been passed up along the way of Evolution by way of the Rotational Energy that was first formed in the Atoms and prior to that by the Electronic form of Energy found in the Electrons and we know now that this electronic form of energy stems from an awareness of self. The electron wants to find its partners and so their life long ambition is to find her, or find him, ones soul mate and pair up with that most significant other for the rest of eternity.

This awareness of Self, is seen at these infinitesimally tiny levels and we can see it even more in the tiniest of all living things, the bacterial forms of life. As we all learn in our first Science classes, bacteria are known to socialize. They use electrical sensors in their sides to learn about who is hanging out in their vicinity and who isn't. They will then clump together in social groups according to their interests in food items and quickly form colonies that can grow into huge clumps that last for centuries some even for millions of years.

We also know that bacteria was the first life form to be created on the Earth. We have even discovered recently that some bacteria actually eat electrons as their main source of food and in so doing conduct electricity from one animal to the next in a long structure of animals that mimics an electric cable. Think about the induction of one state to the next in my above examples.

We also know from studying the DNA molecule the central and most fundamental structure of all life forms known so far that our own DNA is exactly the same as bacterial DNA up to about 70% to 80% of us. This is strong evidence that the desire or need to cooperate with other forms of existence like us is the main property of Consciousness. It's not enough to know that one exists. The next question for the Conscious being is how do I find another one like myself to combine with thus validating my own life along the way.

Now, to cement this in, we consider the DNA molecule in detail which gives further evidence to Consciousness at these atomic and molecular levels. And I want you to consider for a moment how the Monarch Butterflies know where they were born and fly back to their birthplace to give birth to their next generation. I want you to think about how the Salmon know where they were born, after living their lives out in the vast oceans, and return to the same spots in the streams that bore them, if we humans haven't destroyed them, to lay their eggs in the same spot where they were born. Consider how birds know the same spot where they must migrate to escape the winter cold and find something to eat, even though it may be thousands of miles away.

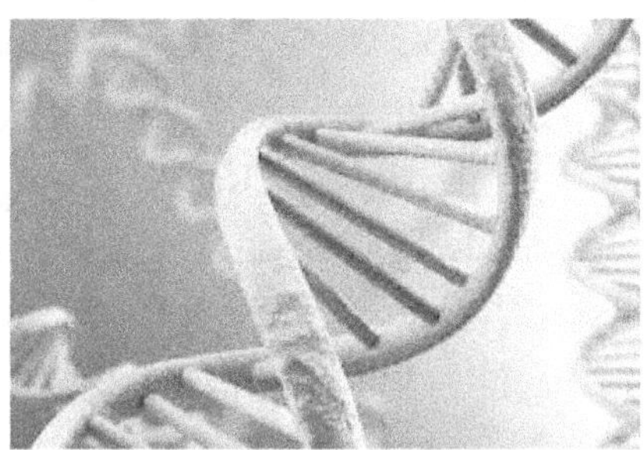

As we all know DNA (pictured above) is the code of life that is circulating around inside the nucleus of every cell of our body. This code is what tells the cells that they are part of a Human Body and all of the detailed structure of your body and mine as well as everyone else who has ever lived or will ever live on this planet.

The first thing we all notice about the DNA molecule is that it looks like a spiral staircase or a long twisted ladder with billions of rungs in the ladder that someone or something should be able to climb and that the legs of the ladder on each side are twisted around each other and this is what is known as the Double Helix or Helical shape of the DNA molecule. The rungs of the ladder are connections made by a simple Hydrogen bond between four components, chemicals that make up the DNA molecule. They are Adenine, Guanine, Cytosine and Thymine. These nucleotides make up a code as they are duplicated all along the rungs of the ladder in what Gene Scientists call the 'Base Pairs'. These four 'molecules' can be combined in infinite combinations throughout the billions of rungs of the ladder.

Genes are large sections of these rungs in the ladder that code for a particular trait. We've all heard that we got our good look 'genes' from our mother and our intelligence genes from our father. This is what they're talking about. The sections of the billions of codes along the rungs of the ladder, through an interpreter called RNA, tells every cell in your body what to do, what actions to take, who to be.

The cells always obey the code and so your heart cells will be made in the quantity required. Your brain cells will be made according to the code, enough to become a brain. Your liver cells will be made out of the proteins that DNA and RNA have ordered to be your liver. All of your muscle cells are made this way as are the cells in your eyes that let you see things, the cells in your ears that let you hear things, the cells in your nose that let you smell things and so on down the line - this is how you are made and how you keep up the maintenance of this magnificent machine for your entire life.

The twisted Helical design of the DNA molecule comes out of the Rotational energy given to it by the atoms that make up the molecular structure. It's just the same way that a wheel is made to turn on the road and give the least amount of resis-tance to the vehicle that is riding on it. The DNA molecule took this shape of the Double Helix because the atoms that make it up have Rotational energy that was transmitted to the shape of the molecule as the various atoms joined up. The Double Helix is just the simplest way to absorb the rotational energy of its construction and maintain its greatest structural integrity. This is Consciousness at work that has been doing the things that it needed to do to create life and which has been hiding in plain sight for billions of years.

The Vibrational State of Consciousness is found in the way that the RNA molecule reads the code and then transfers this information to the proteins by instructing amino acids how to form into proteins that actually do the work of digestion, muscle movement, eyesight, hearing, thinking, running, sitting, making love etc, as part of the trillions of working cells that make up your body and brain. We know it must be a vibration of information because nothing at this level has the ability to communicate in any other way. They are not sending out radio or TV signals or speaking in words or any other language. The RNA molecule actually duplicates your code given by the DNA and then reading these codes, understanding their impacts and then translating these into a different language that proteins can understand and obey.

Is this even more evidence of a microscopic form of Consciousness or could it just be random? The answer should be rather obvious to you by now.

The RNA molecule reads the DNA CODE that is YOU and simply puts out the right energy for the right proteins to be made and they are made right there in the studio where the Consciousness of one molecule whose job it is to take instruction is in tune with the Consciousness of another molecule whose job it is to give instruction. There are several chemical steps in this process that I have summarized, but when molecules get together it's from their Rotational Consciousness and the details of how they do this is actually less important to the fact that they 'Realize' how to do this through this molecular process that originated in the atoms, mostly the Hydrogen, Nitrogen, and Carbon atoms at this stage.

Sometimes in making all of the trillions upon trillions of cells in your body, the Transcription of your code can go wrong. Mistakes are made and this is where diseases can happen in your body. Your code can also be changed by radiation or pollutants in the environment that come out of nowhere, sometimes even from cosmic rays that may have originated in far-off galaxies outside of our own. Our molecular structure, like astronauts on a mission to Mars, know that this can happen and so they are trained to keep the ship safe for the rest of their mission. There are agents in our DNA whose only purpose is to repair the areas in the molecule that have become broken. They are like tiny little machines, the size of a protein, that run up and down the ladder of the DNA helix until they find the 'broken' section of your DNA molecule or mine and then goes about to effect the 'repair'.

NOW, when your DNA strand is 'broken' you don't fall into a heap of useless flesh. No, you don't even notice it at your level of Consciousness, but there's a part of you that does and it knows that it has to go and fix it before things do get bad enough to where you might fall into a lump of useless flesh. How does this repair mechanism 'KNOW' that there is a damaged part of your DNA somewhere amid the billions of base pairs of molecules and then how does it find the one section that needs to be fixed? More quizzical is 'how does it know how to fix the problem? Did part of you go to college and study Genetic Science before you were even born? Perhaps for some of us, but it's not very likely for most.

Somewhere deeper than we can see with our microscopes there exists a vibration of something that is out of harmony with the rest of the DNA Helix and something has evolved inside our DNA to be very sensitive to this vibration and then to know how to make the adjustment in the entire structure so that it sings sweet harmony once again. When we discover this source of the Vibrational State of Consciousness, it will be an important day for Science because it will lead to the solution of some of the larger questions of life and death I am posing here.

So far, we have tallied the 3 states of Consciousness that lead up to us. The 1st state is the Electronic state. Electrons cooperate with other things to form atoms. Atoms remain together and now exhibit the Rotational energy or State of Consciousness. Atoms now cooperate with other atoms of different weights, or numbers of Protons inside them, to form Molecules passing along the Rotational State of Consciousness, but where it must now be called 'Vibrational' because the spring of energy is wound ever tighter with each advance, being that there is simply more and more of it captured inside and therefore gains more momentum.

First the Rotational and then the Vibrational Energy is used for molecules to cooperate to form more complex structures like DNA, Proteins, Amino Acids, Lipids and many others that make up our brain cells, our heart cells, our fingers and toes, our liver and kidneys, or stomachs and arms and legs and lungs. In this process of making life, each molecule made by the Rotational State of Consciousness gradually enters the Vibrational and this 3rd state of Consciousness makes up the far more complex form of life we call Human Life and for this reason our brains become larger and larger and we can use this complex molecular structure every day doing amazing things for our advancement in the heavens like medicine and construction, travel, agriculture, and even exploring the Cosmos or even trying to destroy it.

And, this is what I call the 4th and Final State of Consciousness - Our Own - which I call the Feedback Loop. So named because I find that our highest and best use is to be able to use all of our brain power to look back at where we came from and give someone or some thing out there a nod, a gratitude for making it all happen in such a beautiful, though intricate way. We know all of these things, we are full of so much knowledge because it is our destiny to find it and know it all. And our Consciousness is mainly a Feedback Loop because we use the power of the Electron to gain our highly advanced intelligence and then to use that same power to look back upon the Electron itself and the 1st State of Consciousness and be able to see and judge its importance.

Oh yes, in the same manner that we gain our Consciousness by the inducement on up through the stages of existence, so do we gain the ability to think, the ability to wonder and the ability to know ourselves from the tiniest things that exist - the Electrons - given their magnificent existence by the God Particles.

It's our way of giving our greatest tribute to the Maestro Conductor. It's our standing ovation that we are proud and happy to give in thanks and in wonder, total gratefulness to receive this amazing and unique gift of life.

This may be why so many of us believe in God and why so many countless billions of people have sacrificed everything for their beliefs in God and always will. All of the tiniest things inside us and outside of us owe their existence to the field of energy we know now exists at approximately 125.35 GeV (Giga Electron Volts). We have discovered that it's everywhere and we know it was sent by someone or something that must love us very much to put so much detail into this amazing work of art called the universe.

The only major question left for me is what we will do with this new understanding of ourselves in relation to the rest of the universe. Will we use it for good or evil? Only Time can tell, but I trust in the overall design and how it is this important to God that we exist, so much so that he would construct the universe in this way. It gives me great comfort and confidence in our future.

Chapter Four

- Survival Time

Have you ever wondered at the amazing trait of all living things to strive above all else to remain alive? We call it the 'Will to Survive'. It's the strongest instinct we can find. We all have it. We all study and nurture it throughout our lives. The desire to grow up healthy, the need to send our children to the best schools, the hope of getting a better job, the will to find the right marriage partner are all prime examples of the will to survive. Of course the strongest aspect of it has been muted through the centuries and adapted for a society that has imposed rules of behavior over all of us so that we don't kill, cheat or eat each other in the pursuit of our own survival.

We observe the will to survive in its purest form by looking at animal behavior. Gazelles are the most graceful and beautiful animals ever created, however, Lions are designed to chase them down and kill them and make them their dinner. They feed their cubs on gazelle tissue. Lions could easily kill all the gazelles in their area of the jungle, yet somehow gazelles have existed for millions of years and so have lions. Somehow, the lions know that if they kill all of the gazelles in their neck of the woods, their food supply would run out very quickly and they would all starve. The lion's will to survive has taught them to take only what they need from their pantry every day and the food supply will replenish itself.

Lions take life in order to give life to their offspring.

The same thing is true of any other species except the Human animal. All other species of carnivore does not eat up their food supply greedily hoping to make a profit of it. Somehow they know that if they gobble up all the other animals today, there will not be any left to eat tomorrow. And so, the predators take what they must and leave the rest for another day. In that way, the food supply can keep up with the animals that need to eat them and all species can co-exist on planet Earth for centuries.

For us Humans, it's different. We've used our brain power to conceive ways of managing our food supply by breeding almost every variety of animals and putting them in cages and putting the cages into large farms. This way, we have created a pantry that is almost infinite in size and capacity. We can eat as many chickens, pigs, cows, ducks, quails, ostriches, goats, lambs as we want and never pay the price of the environment running out of food to sustain us because we have captured all the food supply we need for decades and keep them reproducing more and more and more animals that we can then devour. Worse, we believe that we ourselves can procreate to infinite numbers and there will always be this unending food supply no matter how many humans come to live here.

We are already almost 8 billion humans walking the Earth. At this level of population, we are polluting the air and water of our own planet so badly that we only have about ten to twenty years before we suffocate in our own exhaust gases produced by our consumption of the planet. We're using up all of the energy we can produce and turning it into a level of Carbon Dioxide that will suffocate us all and of course all animals need oxygen to live, so we will kill off all animal life on the planet besides ourselves.

Humans are supposed to be more intelligent than any of the so-called 'lower' animals and yet none of the so-called 'lower' animals are consuming resources beyond what they need from the Earth every day to keep their families alive. None of them have a need for air-conditioning. None of them need to drive shiny new cars and trucks around town in order to attract females. No other so-called 'lower' animals are threatening to destroy their home planet. Only the Human animal can claim that dubious distinction.

In the previous chapter of my book, we learned how everything in the universe comes out of nothingness and more or less wills its existence by resonating with the God Particle Field first and then inducing this resonating heart and soul up to the next highest level of existence.

I call this the 'Will To Realize One's Highest And Best Existence'. I put this in capital letters because I think it is superior to the Darwinian notion of Evolution that is known as 'Survival Of The Fittest'. This is certainly a true and accurate way to describe evolution as far as the life of one's own planet is concerned, but how about when you ZOOM OUT to consider the evolution of all life in the universe?

Then, it's not only the fittest animals that survive out there in all of Space/Time. No, now suddenly we're confronted with a desire for all sub-atomic 'life' to become something with more energy and this is what I pinpoint as the origination of the 'Will to Survive' in all animals and even plants that we know about. It's the energy inside us that keeps us moving every day in the direction of our own best chances for survival.

I go into this kind of Cosmic Evolutionary perspective in greater detail in an earlier book of mine 'The Origin Of Creation' wherein, I augment what Darwin proposed to be more accurate. I believe we're here as a living race of beings because we are an example of 'Survival Of The Wisest' and not just the 'Fittest'. For purposes of this discussion, I only want to show you how this look at our 14 billion-year-old Evolution has been expressed in our genes. We'll be delving into this interpretation of Evolution in greater detail later. For now, realize that you are part of the 'Wisest' and not the 'Fittest' form of life. This is obviously true because there are many creatures that are far more fit than we are as humans and they could easily hunt us all down and eat us. Why don't they? Because we were wiser, not necessarily more fit than the lions and tigers that roamed the Earth for eon's before us, and certainly during our reign.

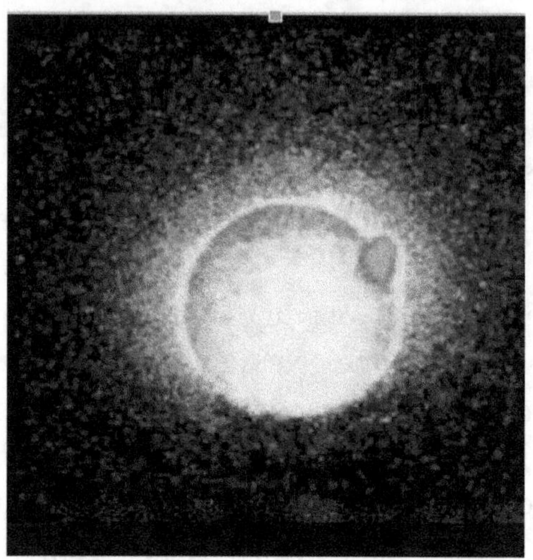

At the moment we are conceived - Shown Here - there is an explosion of light given off from 16 billion Zinc Atoms announcing our birth to the rest of the universe.

The major thought I would like to plant in your brain is that when Electrons and probably all other sub-atomic particles are created, they are created out of their own will to survive. We've already seen that these tiniest of things have a very recognizable form of Consciousness. We have experimented with it and we know it exists and can even measure it. With Consciousness comes a will to survive and this greatest feature we know in the universe originates with the God Particle and it is propagated through the God Particle Field as it spreads out everywhere in the universe. Once something is created, given its awareness of self, it never loses its will to exist. This command to remain in your place in the universe and perform your own individual role, no matter what role that may become, is obeyed above all others made by anyone or any thing else in your life.

In other words, when we feel like we are 'One with the Universe' - it's true - we are. (Remember, I proved this mathematically earlier in this book with my new number - $0_{(I)}$ Zero to the power of Infinity.) And it's ALWAYS true and we never separate from that 'one-ness'. Most of the time, we must suppress the greatest property that all life possesses because we have to to get along in the physical world. We easily put our 'One-ness' inside a place where we can forget about it most of the time because we are simply too busy surviving for the greatest proportion of our lives. No worries, this is going to change someday soon.

We are all united by the will to survive, the major personality trait that is given to everything that we know about. The only reason for this trait to be so important to the Creator is that he or she knew that life would be tough in such an environment where it takes time for the proper things to develop around us. We don't have instant gratification, but we can achieve whatever we want within reason and with tons of patience. When we put this all together, we are unbeatable.

We are 'Universe-Strong' because we are created by the universe in this way but it's from the ground up. Every cell in our body, of which there are trillions upon trillions, each group performing different jobs for our overall machine have inside each of them trillions upon trillions of atoms composed of elec-

trons and protons and quarks that each have the will to survive
strongly embedded in their DNA as well. The will to survive is
like Gorilla Glue except there is no limit to its strength and
power to hold all the pieces of the universe together.

God paints the universe with the broadest brush strokes
first and that brush is filled with 'Will Power' and it will be the
background for any other scenes in the painting. The universe
will survive because it is filled with these instructions at every
tiniest point in Space/Time because this is the first level of
Consciousness.

But the instinct for survival is balanced with the instinct
that the rest of all living things must survive with us in a delicate dance between the needs of the individual and the needs
for all future generations. When a lion kills the gazelle, he
does it for giving life to himself, his pride, his cubs. I believe
these predators know that they are taking a life in order to feed
themselves just as we do as Humans, but even lions and tigers
and other predators pay their respects to their prey whom they
must kill in order to gain that life's energy to be used for their
own. I am convinced that even this sized brain has more respect for their sacrificing creatures than many Humans do. Our
manner of using slaughterhouses, putting the act of destroying
so many millions of animals, our food supply, away from the
consciousness of most people is one of the coldest aspects of
our society.

We think of having a 'nice steak' for dinner, instead of a
slab of muscle from a cow. We think of having some 'Peking
Duck' as if these are visitors from a far away place, instead of

killing and devouring one of the most beautiful animals ever created, probably a mother duck caring for her ducklings who will now starve to death because some cold, senseless person with a shotgun blows her out of the sky and calls it 'Sport'.

No matter how much death we humans pour out into the skies with our shotguns and machine guns, cars, trucks, factories, survival will always remain the biggest part of the universe and the idea of ducks and swans and trees near ponds for all the humble little creatures to live in will stay in place forever. The idea of Humans surviving for all time is not guaranteed as much. Why - you ask? Because we Humans have been granted with a much larger brain that takes up almost half of the energy that we consume. It's a magnificent brain that is powered by all four states of Consciousness. Those of us with Human brains were also given a great responsibility to manage our home planet so that it will live long enough to support many millions of generations of us in the future and all other things as well.

So far, we have not taken our responsibility as managers of this idyllic spot in the universe as seriously as our Creator demands from us. So, we are being challenged and assaulted right now in our Evolution by things that make up our own DNA in order to jolt us into awareness of something much more important to our future than the latest movies, fashions, TV shows or video games. We are at a very unique time in Evolution which could become known as the Fill or Kill era, fulfill our true potential or die.

We must begin to take heed. We have to be in tune with

the Vibrational state of Consciousness and get it boosted up into the elevated realm of our Feedback Loop. We must become sensitive to the needs and the rights of every living thing to survive. We have to do these things by the millions, not just the few. We have to begin a new chapter in our political and economic development where we reward people for doing the right thing for our planet and punish those who do the wrong things to our planet.

We know the difference instinctively and there is no real debate on this topic. If any human activity causes harm to the planet in any way, it's to be punished and discouraged, fined, heavily taxed if necessary until it goes away while the funds generated go into the development of things that REVERSE the harms that we do to this wonderful place - our mother Earth.

This is how we must use our Survival Instinct that was implanted in all of our living cells by the God Particle Field. There will be many others, hopefully some will even be my readers, who will conceive of even better ways to use our Survival Instinct. The dire situation we now find ourselves in will require millions of us acting in concert, in combination and in cooperation as never before on a scale that will seriously impact the topmost priority problem of the day.

If we all attack the problem together, it can be overcome quickly enough to save us. Even if our political leadership brings us to the brink of total disaster, as long as we are ALL UNITED and ready to fight the good fight in harmony with the vibrations in the universe that creates us and holds us in its arms minute by minute, daily, monthly, yearly - we cannot fail.

But, we can only reach this level of power over the forces of evil if we begin to mobilize and activate these energies right now.

This is what I am calling the '4 Stages' of concerted Human effort to save the world because I am envisioning what must happen now as a Rocket Ship launching just in time and headed into outer space to look down upon us and to protect us.

The First Stage is for everyone to look inward for a few days to see if you can feel the Electronic, Rotational and Vibrational energies emanating from the center of the universe and flowing through the land of Zero to the infinite and landing inside your own skull. If you can't feel your direct connection to the source of all energies, please go back to the beginning of this book and try to absorb it once again. If you need more evidence because you want to gain a deeper understanding of this world, then go to some of my previous works that I have pointed to in these pages. In other words, don't take my word for it, do your homework. Look for other citations of this truth. Use the Google Search to find more evidence that supports or rejects this theory of the Cosmos.

The Second Stage of the concerted effort to save our world is one of beginning the co-ordinated effort by multiplying and leveraging the energy released by this book and its effect on your mind. Tell everyone you know about this book and force them to read it. Turn off every device they own until they have completed this study and have surrendered their being to its truth. We won't get past square one until and unless we are all on board, all on the same page, or at least more than a few are.

The Third Stage of the concerted effort to save our world is to get organized about this emergency and start doing things that make your carbon footprint on the Earth smaller and smaller daily. Use mass transit more often. Sell your car, work at home, invent something useful, eat less meat, travel only when it's essential to travel, make home improvements that allow your energy consumption to go down - way down. If you don't have a job that is gentle on the planet find one that is. Look for companies that support the Earth and help reverse the centuries of neglect from the uncaring corporations. Plant a tree. Go on Social Media and shout as loud as you can so that many others hear you, so many others that you know you are making a difference that can be felt by the planet. Make a living record by recording any or all of the things you are doing to turn things around. Put your recordings out there on YouTube, Facebook, Google, Instagram, etc. Don't rest until you have millions of names and faces doing the same.

The Fourth Stage of the concerted effort to save our world is to comprehend fully that it is God's will that we do all things so that we can give back just a little bit to what God himself has put forth in your behalf.

No, it's not difficult to do that. It's actually quite simple, probably the simplest thing you've ever done - especially since you have read this book to this page. You have all the information you need to know how much God has put forth in your behalf. It's really not Rocket Science after all, is it?

I think that this is a good place to make a note about - 'Creationism'.

There are people who go to Church and are told by their pastors that the Earth is no older than 7,000 years. They ignore all scientific evidence of building and early human settlements that go back 100,000 years or more. They ignore the early humans who lived in caves and left paintings and pottery and other evidence of their existence at this time. They say that Dinosaur bones were made for the movies and some of them did live 7,000 years ago and that Noah even had Dinosaurs on the Ark that saved God's chosen few when he made it rain so much that the Earth was under water, of which there is no evidence. Oh God! I could go on and on about this ignorance and atrophy of the Human mind.

They want you to believe this only because it is claimed in the Bible by some unknown goat and sheep-herder who decided that God created the Earth around this time, 7,000 years ago. It seemed long enough to him. It's so ridiculous a belief it breaks my heart that there is anyone over the age of 10 who will believe these fakers. Yet, sadly there are millions of people who have abandoned their brains and common sense and will believe this horse manure simply because a Priest or Pastor told them this nonsense and ordered them to believe it, and why so that they will continue to be mentally dominated by this cabal and continue contributing their time, money and energy to the Church.

But, the most frustrating part of this over-zealous and primitive 'Religionism' is that the people involved are merely wanting to believe in God. This is the greatest instinct that I have been talking about that is inside all of us. It's only misdirected in this case. God does exist and if these over-zealous people would simply read a few books about Science and Spirituality like this one, they might find a better way to dedicate their lives. These people are not stupid. They are only easily deceived and duped by people who are really skilled at deception and 'duping'. We can only live in hope that they will finally turn their brains back on and come here to learn the Truth.

.

The Truth, as you my readers have just learned is that God does exist, but He or She doesn't need to be praised, doesn't need to make us prove our faithfulness, and God certainly doesn't need any pastors or priests all dressed in robes to make themselves look the part. God exists in all things. He or She directs all of the largest and the smallest bits to act in harmony in support of us, and all God expects of us is to want to know more.

You now know much more now than you have ever known, if you've read this far, and you have fulfilled everything God needs from you. You now know more about God than all of the religious fanatics from all of the planet's religions have known for thousands of years. You have more truth now inside your brain than the combined brain power of millions of priests and super-fans of God's since the beginning of time up until the present. And no, God doesn't need you to go to Church to study someone's fairy tales and dogma. And, in all consideration of the facts, God prefers you refrain from such activities, because it's time we all grew up. He or She doesn't need you to pay him money every week. God doesn't need you to pray for stuff either because he gives you exactly what you need and no more every minute of your life, nor does He put problems in your way that you cannot handle.

All I do is give thanks to God for all of these things that God has allowed me to see and try my best to give a little consideration back as part of my duty as the 'Feedback Loop', the 4th and highest state of Consciousness as you have learned.

Now, there is the other side to deal with - the Atheists.

There are actually people who are on the opposite end of the spectrum from the religious fanatics. They are fanatical that God does NOT exist. They call this syndrome - 'Atheism', which literally means - 'No God' - none. We're alone. Everything you see in the universe is random. Nothing has any meaning for this sad lot. Consciousness is just something that we have evolved and sadly there is no reason that it happens.

To these atrophied brains I make the following arguments.

More and more study forces us into the realization that randomness in the architecture of this universe is out of the question and people who continually charge randomness in the construction of the universe are ignoring all of the dovetail joints. The building blocks of the universe all show us how they were constructed with the notion of them snapping together, easily and uniformly. Some knowledge outside of the parts of the universe had to design them to snap together in such an easy and simple way that cannot possibly be random or chance events, as many scientists will argue.

Some Presidents make sure they get the country's highest honor - The Medal of Freedom deservedly or NOT!

The son of President Ronald Reagan, Ronnie Reagan, appears on Television every once in a while asking us to support his group of Atheists. Apparently, they need money to spread the word of nothingness, no meaning to our lives, no design in the universe. I have no idea why someone would want to tout his or her belief in NO God, nothing outside of ourselves that we can look up to, pray to, ask for help to, give thanks to. What would be the motivation of someone like Ronnie Reagan who seems to have an overwhelming urge to dissuade those of us who want to believe in a guiding hand, a force that is the source of all of our beautiful and life-giving orders of existence here on this little dust ball spinning so quietly in Space/Time? Why would someone want to break us of our desire to see the world in a more poetic way than someone who just wants to ignore everything that is so wonderfully and graciously designed for our well-being?

Why doesn't someone filled with 'Nothingness' and 'Meaninglessness' just keep it to themselves and instead try so diligently to take away our joy? What a rotten bunch of apples these guys are. Why would anyone want to spoil our view of the universe, much more magnificent than his? Is it jealousy? Do the Atheists think that they are smarter than us?

I have friends who are Atheists, not even agnostics, but pure and avid Atheists just like Ronnie Reagan and his ilk. There are millions of them. Whenever I get into a debate with these people, and only after I give them some cross-section of the evidence that I have been working with, I ask them - 'So you think that your little brain is capable of understanding the entire universe and conclude that there is no guiding force? You are smart enough to rule out this possibility even with all the evidence that YOU CAN SEE with your own eyes that teaches otherwise?' Or are you blind?

They never have an answer to this question of mine and sometimes I can even see the wheels starting to turn inside their heads so that the next time we have this debates, they have altered their arguments just slightly to allow for more possibilities of my theory I want to label as Scientific Spirituality.

As you know, I'm also making a claim here in this book that the proof that God exists - as in a guiding force or set of principles, a Cosmic Consciousness - lies in the fact that all living things and even in observed non-living things, and far above all other instincts is the 'Survival Instinct' as it is handed up to us by the sub-atomic, atomic and molecular forces. And so Evolution is not simply the 'Survival of the Fittest' but that it's more true to say that our Evolution is all about the 'Survival of the smartest or the wisest.' And only the strongest in this sense will survive from this day going forward because only the wisest among us will be able to rise to the challenges that we have put in front of us. The problem today is ourselves, too much Human development and it needs to be curtailed and channeled in a completely new direction.

Now, the proof of this new way of looking at Evolution will come in the next few years when the so-called 'Fittest' of us, the bigger humans, the fastest, the most athletic, the most heavily muscled, even the most politically powerful individuals will be marginalized and circumvented in order for the rest of us - the 'Council of the Wise' as I like to call us - the vast majority of us, band together in ways as yet unknown and we begin to plan and strategize our way out of this mess we've gotten ourselves into.

The most physically fit of us will be the last to join us and they will do so only because it will be their last chance to save themselves and it will be bleedingly obvious even to this classification of us that they are clearly not the top of the jungle any more and that they may have missed something and will start to listen to us, instead of us always forced to listen to their meaningless clap-trap about all things insignificant and small.

AND it's NOW or NEVER that we must realize some of these things and take them into our hearts and begin a new Race of Human Being. Being Human, we have to quickly and suddenly realize our origins and our fullest potential, our highest and best use. And, this will be the subject of the next and final chapter in this book.

We have the wisdom and creativity today to make - as Neil Armstrong said, 'One small step for a Human and one giant leap for all Mankind' but not the kind of leap that got us to the Moon. We need to make Spiritual Leap that takes us into the heart and soul of the universe and gives us all the energies to make our form of Consciousness the greatest of God's great achievements. We have to begin to think as ONE. We need to organize our priorities as ONE. We need to make the Earth our true home and sacred temple to Life. We need to begin to give what we were intended to become as ONE. We need to be the Feedback Loop and show our Creator that we now understand it all and want to give thanks in the only way that will be the most meaningful and will register as the loudest in the overall scheme of things.

We need to become true Stewards of the Earth. We can and must do this and in the next chapter, I will give you what I feel is our best chance at using all our brains to get it done once and for all and before it's too late.

Chapter Five
- Virtual Reality

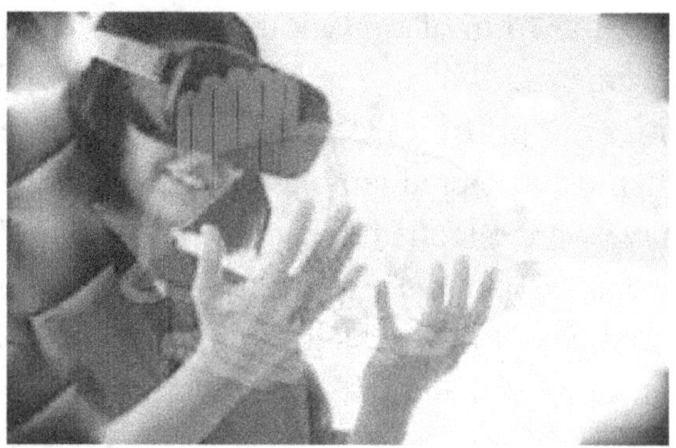

When you see someone wearing a helmet with a viewfinder that only they can see into, you're probably looking at someone using the newest computer craze - Virtual Reality. In this realm, clever computer graphics designers and game designers take us into a new kind of video game that is so vivid and uses 3-D graphics so cleverly that we think of ourselves as being in a new and exciting type of Reality. And some are ready to think of this very primitive first step into Virtual Reality as a 'better' place to live out their lives than the Reality that you and I have become accustomed to living in and which has lasted for centuries.

This will engender a Technological Revolution unlike any other in history. It will dwarf even the invention of the computer or the invention of the Internet combined. How can I say that?

Let's look at just one example of what is in store for us assuming that we can survive the next few years. The following are examples of how I believe we will be using this new technology for our own good once we get over the hump of the current selfish rush for more and more consumption of the Earth's resources.

Without the awkward and clumsy 3-D helmets, we will be able to see our offices without stepping out of our own homes. And, other workers will be able to see us interacting with them in the actual work environment in similar ways that our actual bodies would interact.

AND - our interactions will feel the same as before. In other words, we will be able to work together as we do now, talking to each other and even working side by side with tactile responses - in other words, touching each other as we do now. We will know we're at work because the working environment will be the same or better than it is now for all of us who choose to work. When we go on vacation, the same complete experience of being somewhere will be ours to enjoy without leaving the comfort of home. We will be able to place our bodies in a nice comfortable chair in our back yards or in our living rooms and suddenly be transported to Hawaii, Fiji, Paris, London, Rome, New York, Los Angeles, Morocco, Singapore, Tokyo or even Mars or Ceti-Alpha 12.

The use of fossil fuels will be nearly Zero or less than Zero because we will be getting all of our power from sustainable and renewable and clean sources like Solar, Wind and Water.

Better than entertaining ourselves, we'll also be able to learn and teach one another about all of these details of how the universe is constructed even to a depth I can only imagine. We'll find ourselves at the nucleus of the Hydrogen atom and looking up into the electron sky. We'll hear the humming of the energy fields colliding with and reflecting one another back and forth billions of times a second while creating the background wave action that we know up here in the upper levels of the woof and warp of what we have learned is the fabric of Space/Time. We'll see, taste, hear and feel, along with many other sensory responses, unknown to us now while we play our little part in the hive and be informed by it.

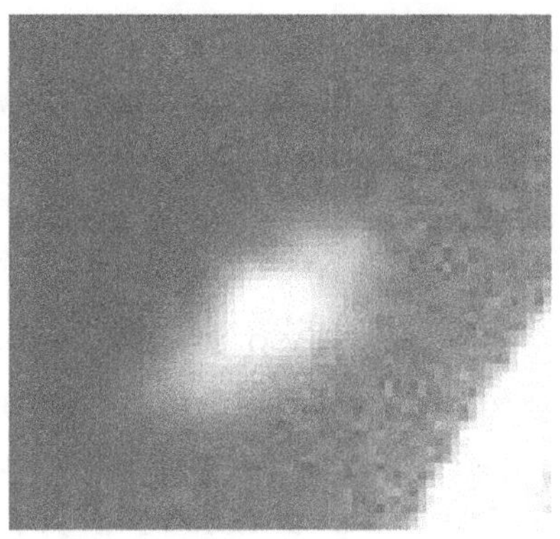

An Atom hovering in space.

We will help in the Creation of it as it is formed. We'll find ourselves in the orchestra pit watching the Maestro giving us the 'Down beat'. We'll take our instructions from him and he'll even provide the instruments for anyone who wants to play along. We'll know the particular place where we come in and we'll come in expertly and smoothly and in perfect harmony and it will be thrilling beyond description.

We'll be able to reach new heights in all areas of endeavor. Our travels through Space will be faster and better because we will have simulators in Virtual Reality for nearly every possible experience which is a form of training, but could also end up leading to another advance in technology that revolutionizes Space Travel.

We will be able to learn how the biology of our own body works to a degree never thought possible as medical technicians are able to explore how diseases are propagated around our bodies through the blood vessels. We'll be able to invent far faster and more thoughtful machines as we are able to trace the electrical signals throughout any type of system. We'll be able to invent new technological miracles at a pace that we've never known before.

With Virtual Reality of such an advanced nature we'll be able to achieve the highest level of Consciousness the universe is capable of providing. We'll evolve new systems of governance that is fair and equitable to all segments of society, young, old, skilled, unskilled, educated, uneducated. We'll be able to handle any emergencies of any proportion because we will be able to predict more accurately which ones are going to hit and when. We'll evolve systems and methods to tackle them all before they rage out of control. In short, we can have the ideal civilization that we Humans can truly boast about and export all over the galaxy without fear of being bested by some more advanced race.

This all assumes of course that we survive the next ten to twenty years of incompetence, extreme carelessness and neglect. Our Industry and Government leaders knew, for example that the heavy use of fossil fuels would someday be a serious threat to our entire planet by pouring far too much carbon into the atmosphere overwhelming the Earth's natural systems that always tend to balance out these emissions through the respiration of the forests and the oceans. And of course, in the meantime they killed off the forests and destroyed the oceans.

They hid this information from the rest of us because if they were thwarted in spreading their toxic Industry all over the world, it would cost them huge profits. Being the greediest and most selfish animals in existence today, they fought all efforts from those of us who love the Earth and want to remain on it for a much longer time than they have planned for us.

And yet, through God's good graces, there is hope, though it is a modest slender ray of hope, just peeking out of the clouds of Doom and it is brought to us at the last minute by the tiniest of biological units on the planet. Through the highest and best use of technology we have mustered so far, we may be able to Mac-Giver some emergency systems to guide us out of the mess they produced for us with all due diligence and speed. It has to be quick and that means it will not be pretty.

We will make mistakes that will have to be corrected quickly. We will do some things that are not as effective as we like. But, we must hope to learn from our mistakes quickly and then redouble our efforts to clean up and reverse the carbon emissions problem that we now realize will bring us to total ex-tinction very soon.

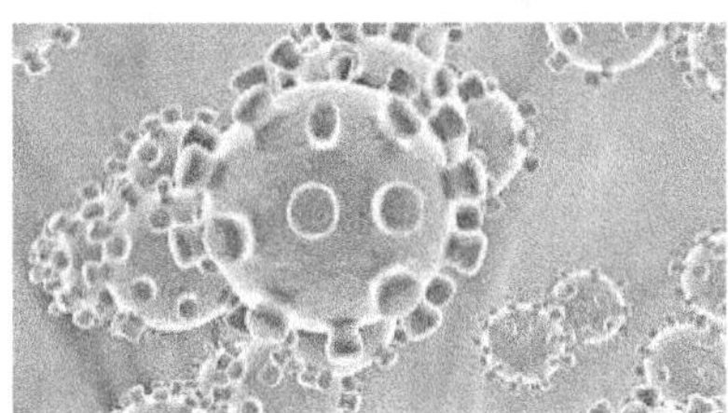

The Novel Corona Virus - Covid-19

The current Global Corona Virus Pandemic is an opportu-nity that we must exploit. Because of the attack of this virus, we have all been forced to stay at home and take shelter for months now. It could last for a few years off and on. Because of us not going to our jobs every day, the amount of soot and toxic fumes we are pouring out into the atmosphere of our planet is negligible and can be neutralized by the existing trees and ocean plankton that is still desperately hanging on for their own lives.

There's been much speculation about where the 'Novel' Coronavirus - AKA Covid-19 - was born. Some have said it was started in a market in China where they keep live bats, dogs, cats and other little innocent animals on hand until some-one wants to eat one and then they slaughter it in front of the customer and he or she takes it home to put into some ghastly medieval stew. They say this because the first known cases were traced to this market in Wuhan. Others have noted that there is a Chinese military biological warfare laboratory within a few hundred feet of this market and that this virus was cre-ated by Chinese military for use against an enemy in war.

No matter who or what is ultimately found responsible for this, virus DNA has actually been around for millions if not bil-lions of years. This very basic form of life found a way to replicate itself very fast and over the 5 billion years of Evolu-tion of these simple life forms, their DNA has been carried for-ward by this form of replication that has become a major part of our Human DNA.

More than half of our own structure is Virus DNA and Bacterial DNA, two competing forms of life that were the first forms of life on Planet Earth. As the first forms of life, the will to survive that is embedded in the virus and bacterial code of life and given by their own sub-atomic will to survive, as we have discussed, may be stronger than our own. The virus and bacteria's will to survive has been reinforced by billions of years of existence on planet Earth, whereas Human will to survive has only been existent on this planet for a few million years. The current invasion of this little snippet of our own DNA could be a warning from a life form that is determined to stay around even if we are gone and therefore this could be our greatest WAKE-UP CALL and perhaps our last one as well.

The warning is coming from inside of us. Something very deep inside of us is trying to tell us that we only have a few years left - and they could be the final curtain call for all of the Human Race if we do nothing to avert the upcoming disaster. This means that our own basis of Consciousness that has brought us to this point of being the dominant creature on the Earth is now trying to pull us back from the brink. We must heed the warning. I have no doubt that we will heed this warning, but I'm afraid that we may not take significant action before millions of us are sacrificed to our own stupidity and apathy. Inertia is a strong force and so the Industrial inertia of using fossil fuels could be stronger than our will to survive as a species. I hope not.

We can always replace lost jobs and lost income from the ending of the fossil fuel era, and set the economy into the right direction but we cannot freeze-dry a few million people, store their bodies in a vault somewhere that will preserve them long enough and wait for the Earth to come back to life and then thaw them out like so much meat and start civilization over from there. This may be what some in the fossil fuel industry have planned for us, but I'm not sure this is a solution that will work for the vast majority of us.

And who gets to choose which individuals among us are worthy to be frozen and stored, perhaps for centuries, for some distant uncertain future? We could of course all donate some blood to a blood bank and hope that some future life form will be able to extract our individual DNA and bring us all back to life in the future world by cloning us. But, why would a future intelligent race of creatures want to bring back the creatures, or even the species who may have destroyed all life on the planet? It's much more likely that some future advanced life form would destroy the blood bank where our DNA is stored and this would be the first thing they do if they have any brains at all and want to protect their own future. Why would we bring back the Dinosaurs, unless it was for some entertainment park? We certainly would not want them walking the streets.

No, this is our final warning and we must start making small sacrifices today so that we don't have to make the ultimate sacrifice of billions of people alive today or soon to be born into this mess.

I propose that we start using and improving our technological prowess in all areas but especially in Virtual Reality to educate and inform the people of our planet to WAKE UP and show them ways to avert the coming disaster in changing their own lives in 'teensie-weensi' ways that can cascade into bigger and bigger activities that can collectively reverse the Earth's warming and pollution.

Through Virtual Reality and related technologies, let's start developing a Virtual work-place where we can all go to get our regular jobs done without ever having to use fossil fuels to do them, or at least in ways that greatly reduces the use of the carbon dioxide and other green house gases. We know now, through the activities of this virus that we can use distance learning, using virtual classrooms instead of the old model of students sitting around in an actual classrooms. We have seen government work by limiting the amount of face time given to citizens served. Most of us can get our jobs done remotely so why are we all driving life-killing machines to work every day?

Why aren't millions of us going to work to fabricate things that will replace the carbon-emitting machines? Very soon, the vast majority of us are going to have to start pulling in the same direction. Why not us? Why not now?

For those of you who are asking how we will finance such a huge change in our economy - I have one word for you -

Crypto-Currency! Something like Bitcoin, except a Bitcoin that is generated to a greater or lesser extent by activities that support LIFE - all life on the Earth instead of the currency we use today that supports the activities of Death to all life on the Earth.

The choice is obvious to me. How about you?

In Summation:

If you've been following along with the main theme of
this book about the 'Science' of Physics, you may have
realized that everything in the universe is dependent upon the
smallest pieces of the universal jig-saw puzzle. It all snaps
together to form the universe we live in, the planets, the food,
the energy, the people, all the things that make everything work
so beautifully together. Every nook and cranny of the universe

is composed of a granular layer of the smallest sub-atomic energies that we call particles, but aren't really. This is just how we can imagine them better.

The point is all of this energy at this level starts to spin from the resonating force of the interaction between all things and the God Particle Field. This resonating 'kinetic energy' that makes up every little particle in the universe is passed all the way up to the largest structures in the universe. This process of course would have to include your own physical presence.

And the proof of what I just stated is the fact that everything in the universe is involved in some kind of 'spin'. From the smallest things to the largest things, the stars, the planets, the galaxies, clusters of galaxies and even the black holes and the darkest unknown regions of Space/Time. It's all spinning.

Mostly remember that the 'Spin' is how Consciousness is passed up the 'food chain' from the tiniest things in the universe and induced up to all the largest things that we know, ourselves included.

So, all you have to do to make something out of nothing is to remember this little phrase of mine - "The Spin I'm In'. When you listen to music, communicate with another person, travel to a destination, eat your meals, drink your beer or wine, whisper this little phrase to yourself and you will start to perceive the universe the way it really is and/or how you want it to be. Sometimes this brings on great and wonderful results and sometimes, this 'real-time realization' will bring on doubt and maybe even great disappointment. But, realize too that even this is 'Spin'. It's your own spin that you're putting on things and it could be the wrong spin.

This means you just have to keep your options open and continue monitoring the spin until you feel your own spin coming into a perfect synchronicity with all the other spin around you. Now, you're cooking. You'll begin to hear the great orchestral concert being conducted by the Maestro himself or herself. You'll eventually find your own instrument and suddenly know how to play it in perfect harmony and the great Arpeggio continues to brighten with the addition of your own music to the score.

Which reminds me of an old Johnny Mercer song.

'Baby, down and down I go
All around I go, in a spin
Loving the spin I'm in
Under that old black magic called love'

Get out there and get involved in the spin of the Old Black Magic called Universal Love.

AND WITH ALL THE KNOWLEDGE we have attained over the last few years - WE CAN DO ANYTHING.

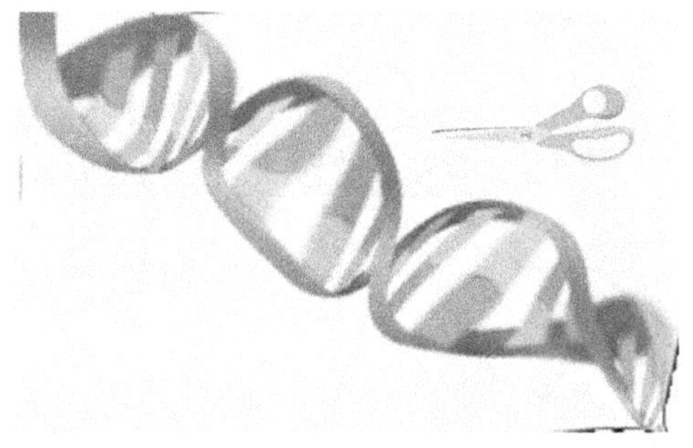

With Gene Editing technology of today, we can be anything we want to be.

Let's remake Humans.

# # #

Conclusion

A large portion of the Scientific Community will not take kindly to the theme of this book, at least not at first. Perhaps a few years or maybe even a few decades from now, further evidence will start to be uncovered. So far, we have two Black Holes producing the mysterious whistle in Space discovered by LIGO on the day that they turned it on. We have the God Particles, the very central piece of the Standard Model, most generally accepted version of Quantum Physics, actual messengers from God, confirmed to exist by the Large Hadron Collider in Switzerland in 2012. We have the Zinc Spark of life that explodes a microscopic, yet gigantic fireworks show in the womb when all of us are conceived.

We have even discovered that Electrons and probably all sub-atomic particles have Consciousness, in that they are at least aware of one another's physical state and position in the universe and are able to communicate over distances that may seem huge to us, but are a Zero distance to them.

We have the miracle discovery in our own lifetime of our own Code of Life, our DNA, a string of four nucleotides that forms a code that is interpreted by every single one of the cells in our bodies and instructs them what part to play in the symphony of molecular activity that is us.

Today, we now have all these major discoveries and on top of all that we have my new formula that shows how all the forces in the universe are united into ONE, something that Albert Einstein believed to be true, but could not prove. Today, we have that proof.

$$0^{(i)} = U$$

Zero to the power of Infinity = Any Number –

You can think of this formula as an I.O.U between yourself and the universe from where you have descended.

In spite of all this recent evidence, the Scientific Community will scoff at my ideas for one reason only. They always tell me when we get discussing the spiritual aspect of our universe that none of it can be tested. And, to that extent, they're right. You cannot have a final test for any of the theories that I am putting forth in this book because they are too big as concepts and will never fit into a small enough test tube or any set of instruments devised by the Human brain. Cosmic Consciousness is far beyond any Human Testing skills and probably always will be.

And, besides the practical aspects of this testing, it's not the purpose or intent of Consciousness that it requires us to know and respect it, as the God in the Bible seems to require us to do. It's been put there for everyone, but not everyone needs to appreciate it and there is no section of our DNA that forces us to believe anything. There are no genes, at least none that have been discovered to date, that code for the need to know our Creator. As I'm writing these words, however, I'm aware that someday soon, some genetic researcher somewhere may find that section or sections of our DNA, that I am calling the God Gene that does give a person a predilection for the Spiritual things in life, in other words someone like me and perhaps you too.

Lacking this gene, a Human Being will appear to be cold, unsympathetic, self-centered, sometimes even cruel or hostile. Perhaps you have known people like this. Don't hate them because it's entirely possible that these cold uncaring, greed-filled anti-social people walking this Earth were simply born without the God Gene. They have no moral compass. They are constantly tormented about the meaning of life and the futility of their own life. They won't show it, never opening up with anything honest or genuine about themselves and will always have that Stoic look plastered on their faces. You don't want to be like them.

Now, that I've covered so much of the historic basis for our existence in this book, I can almost feel it in my bones that I may have the God gene. If I have it, I think it's safe to say that many millions or even billions of Human Beings have this gene. Do you feel that you may have it? If so, pass it on. It may be our only hope.

Indeed, the God gene may be why so many billions of souls over the centuries have fallen into this kind of hope for and belief in a superior being, a Creator of everything we see and know. And, if it is discovered one day that the tendency of a Human Being to believe in God or not resides in our DNA, mostly, I wonder if that will be enough proof to the Scientists among us when they are forced to realize that their own objectivity and love of a 'Proof of Concept' is due to the way that they were designed in the first place. Because we would all have to conclude that if there is such a gene in the Human Genome that the Creator of all things would have to be the one to put it there.

It's only been a few years since we Humans have discovered and completely mapped the Human Genome. This is like saying that it's only been a few years since Christopher Columbus discovered America. Imagine all of the amazing events in the last five centuries since this pivotal turning point in our history. The same thing, and even more so could be happening now in the years ahead when we have discovered a completely foreign realm of knowledge and the pioneers are just now arriving on its shoreline. Imagine the great discoveries that are yet to come now that we know exactly what we're made of and where it comes from.

Just as it took 50 years from the time that Professor Higgs and other great physicists predicted that there had to be a God Particle, something that brings all the other particles into existence, perhaps it may take 5 decades, but hopefully not much longer when they discover my theory is right and they will be able to isolate the genetic tendency to believe firmly in a Creative Force, one that actually has plans for us, and gives us just enough Space and Time so that we can know and pursue our true destiny. This to me is the definition of happiness.

I can only hope that I live long enough to know the proof and finally I can say to the doubters - 'There you go. There's your freaking test!'

AND one truly final thought.

A few years AGO, a dear friend of mine, Professor Jean Brodie invited me to attend her lecture series on Astro-Physics at the University of California/Santa Cruz. It was the most interesting and compelling course I've ever taken in my life.

At the end of her final lecture, she had a picture of the Big Bang up on the Screen and she said:

"At the first trillionth of a trillionth of the first second of The Big Bang, all the information that the universe needed was imparted."

I raised my hand and I said: "Jean, did you mean to use the word 'Information' and she said 'Yes' and finished the rest of her lecture. Later that day and for many years after this lecture of hers, I've considered that statement and how this one fact of Astro-Physics research into the heavens supports of all my suspicions and crazy theories.

Because if you think about it - for there to be a need to create INFORMATION that the universe uses to form itself to the point where we enjoy it today - there has to be someone who the Creator of all that information is pointing it toward. That can only be us. And until and unless we find other forms of life out there, better, or smarter than us, then - it is us for whom all the information is put forth - for us to discover it and give thanks.

I raised my hand and I said: "Jean, did you mean to use the word 'Information' and she said 'Yes' and finished the rest of her lecture. Later that day and for many years after this lecture of hers, I've considered that statement and how this one great discovery of Astro-Physics research into the heavens supports of all my suspicions and crazy theories.

Because if you think about it - for there to be a need to create INFORMATION that the universe uses to form itself to the point where we enjoy it today - there has to be someone who the Creator of all that information is pointing it toward. That can only be us. And until and unless we find other forms of life out there, better, or smarter than us, then - it is us for whom all the information is put forth - for us to discover it and give thanks to whoever or whatever put it out there.

About The Author

Michael Mathiesen

Someday, I would like to be known as the discover or the 4 States of Consciousness as well as the Three Laws of Consciousness

You'll find out why if you read my two books with the same titles found below.

If you want an Audible Version of this book and/or most of my previous works - please visit:

MyAudibleBooks.com

If you got anything out of this book - please give it a good review wherever your found ut, but only if it makes a difference in your life and that you have learned a thing or two. The more reviews the book gets, the more Amazon will show it to other readers. And this book sorely needs to be discovered by millions of people if we are going to make a big enough difference to save the planet. Everyone has to play a part.

IF you also like Classic Science Fiction – AND based on my earlier ideas found in this book...

Try my new book:

My Cosmic Brain And The 2nd Big Bang

OH – YES! There's one coming!

MyCosmicBrain.com

This book is all about how they took my brain and placed it inside a Scientific experiment - the Think Tank and ran some tests. What turns out blew their minds and I believe it will blow yours as well. Oh, it's also a Love story.

Prior to this book, I wrote

Zentanglements –

The Three Laws of Consciousness for Smarties

Zentanglements.com

Zentanglements are a new way of forming images that tie into an aspect of the Entangled Universe that I talk so much about. For this book, I produced a number of paintings and drawings that inspired me to glean more information out of our amazing universe.

My Illustrations can be found at:

CafePress.com/MyFineArt

Prior to this book, I also wrote - Extinction : Live

This is taken from the idea that Elon Musk's SpaceX is now planning to colonize Mars. I wanted to support that idea by helping to show the world HOW to colonize another planet in such a way that what they learn on MARS could be used on the Earth to REPOPULATE and Re-TerraForm our planet.

www.Extinction.Live

Watch the Extinction of the Human Race LIVE as it happens.

AUDIBLE BOOK Available at Audible.com

- - -

Just prior to this book - I authored:

Audible Book

We're living in times of miraculous discoveries and potential for the Human Spirit has never been greater - but are we getting the biggest bang for our bucks?

You can TAKE THE COURSE at:

Udemy.com/Course/Miracles

You can also find almost ALL of my books in the

MOST OF MY BOOKS are ALSO FOUND ON AUDIBLE As well as Print and Kindle

MyAudibleBooks.com

Prior to this book I Authored:

The Audible Book

Print And Kindle Version

I was born in Boston, Massachusetts at an early age!
(Dumb, I know.)

I remember visiting the site of the Boston Tea Party. And I especially remember visiting the Boston Museum of Science which obviously had a great impact on me - although it was little known at the time.

I grew up in a beautiful little town, Natick, Massachusetts not very far from the first battle of the American Revolution and the 'Shot Heard Round The World' It was an idyllic place that I have often likened to 'Walden's Pond'. The place where I learned everything about Nature.

I have always felt close to the people who founded this great nation - but over my life-time learned that they were all slave-owners and then I started to realize that Slavery still exists today. So, I wrote a book about a new kind of America that may someday live up to all the ideals that we've fought and died for over the years.

AND in my most recent book - I have delivered on the way to FREE all the SLAVES - all of the American People.

through - 'REAL DEMOCRACY'

It's got to happen someday soon.

They call the United States a Democracy, but absolutely no one
is ever allowed to vote for anything. NOT YET - NOT UNTIL
WE ALL DEMAND the 28th Amendment

Prior to Real Elections, I wrote:
Because it's not enough that only ONE COUNTRY ENJOYS
a REAL DEMOCRACY - UNTIL and UNLESS the WORLD
ENJOYS ALL THE FREEDOM THAT GOD HAS
GRANTED US - There will be no HOPE for us as a
TRUE and LASTING CIVILIZATION!

The Audible Book

The Print Version

Prior to this book - I wrote:

Audible Book

Print Version on Amazon

This book is all about how to UPGRADE our Electrical Grid so that we can STOP and then REVERSE the Climate Emergency IN TIME TO SAVE THE PLANET.

This is a totally innovative plan that I devised where we combine all of the RENEWABLE ENERGY SOURCES in ONE FELL SWOOP in a completely new and modernized GRID.

Prior to this book - I AUTHORED

The Audible Book

Print and Kindle Version

Please go to AMAZON - search for these books and then SCROLL DOWN on the sales page where you can add a few sentences and give it 5-stars - Thank You.

YOU REALLY NEED TO HAVE BOTH BOOKS
WE NEED YOUR 5-STAR REVIEWS BECAUSE there will certainly be an ARMY OF SCALLIWAGS who are fearful of change AND/OR firmly believe that the Human Race should be eliminated.

www.GreenNewDealOfficial.com

Take the Online Course

GET THE AUDIBLE BOOK

bit.ly/GreenNewDealAudible

Please tell everyone you know about this book because we all need to FIRST LEVERAGE ourselves in the most effective way. Then, doing everything else that we mention you can be doing in the final chapter. BUT FIRST - if this book is not a RAGING BEST-SELLER - we may not survive the coming catastrophe.

THEREFORE, send everyone to our Amazon Page or our website where you can gain access to news and updates regarding SAVING OUR EARTH plus access to all my other books.

www.GreenNewDealOfficial.com

ALSO go there often to learn about updates to this book - new information, new challenges, new things that YOU CAN DO.

Please get either the PRINT or AUDIBLE BOOK TODAY For even more Revolutionary new information you need to know and share with others.

PUT THE LINK GreenNeDealOfficial.com in everything you do - all your emails, IM's, Facebook posts, Tweets, Etc.

*   *   *

Prior to this book - I authored a Science Fiction Thriller based on many of the most exciting new breakthroughs in Science, such as the fact that many of our sub-atomic particles become entangled with one another, which means they know about the existence of their life-long partner - the first indication of soul mates?

Also, I have noticed that the situation we're facing today in many ways is the opposite of what we should expect in our Evolution by this point in time. It's a BIZARRO UNIVERSE and I think that there was a cross-over point, a nexis in the path of two or more time-lines and somehow we crossed over into the wrong path for humanity. I put it down to a cloud of ANTI-MATTER that our planet may have travelled through recently or may still be travelling through it.

I hope you'll give it a read.

AND SPREAD THE WORD - IF everyone has this information and SHARES IT - WE'RE GOING TO BE OK~!

The Audible Book

PRINT and KINDLE VERSIONS
bit.ly/brain-drain-book

**Science Fiction based on the life and work of
ALBERT EINSTEIN**

What if they took Einstein's brain after he died and kept it alive for decades? Imagine the ideas he could have given us in perpetuity. If you like 'Thought-Experiments' of Einstein's this book is for you. Classic Science Fiction.

You can find other titles of mine on Amazon

Some of these are:

- Real Elections 2020

- **Only You Can Help Save The Earth**

- The Green New Deal
- - The Rise of Democratic Socialism
- Brain Drain
- Someone Completely New
- Autonomous Government
- The Future of Cars
- **Anti-Matter (Science Fiction)**
- **The 4 States Of Consciousness**
- Can Electrons Learn?
- Maxtricity - The Science of Entangled Electrons
- The Force Equation
- One Crazy Bastard Saves The World (Sci-Fi)
- The Origin Of Creation
- The Flux Antenna
- Metamorphosis - #AICocoon
- The Doomsday Watch
- The God Particle Bible

All of these titles are also available on Audible.com

Prior to this book, I authored - Autonomous Government and I authored the Science Fiction novel - Anti-Matter, as well as several others I list for you below. These previous works were my first inspiration for the ideas in this book. I hope you'll give them all a read.

It's BLEEDINGLY OBVIOUS to most of us that HUMANS HAVE FAILED US - Perhaps it's time to let ROBOTS TAKE OVER.

The Audible Book

The Print and Kindle Version

Prior to this book - I authored.

The AudibleBook

Print and Kindle Versions

This book was also inspired while writing about the Future of Cars because the same technology of autonomous cars can be simply and quickly adapted to provide the mechanisms for the Autonomous Government.

The Future of Cars was inspired by this book, published a few weeks prior - Anti-Matter

The Audible Book

The Print and Kindle Versions

When things go bad in our lives, we usually put it down to just 'Bad Luck' or 'Karma' or 'Fate'. It's none of these things - It's the influence of Anti-Matter in our vicinity of Space/Time.

Another Classic Science Fiction story of mine.

GO TO

TELL YOUR FRIENDS - About this new discovery - Send them this URL

http://bit.ly/greataudiblebooks

Or just search in Amazon.com for the Print and Kindle versions

This book, though fictional for the most part, possesses elements of Science that I believe are hundreds of years ahead of generally accepted 'truths' of Science. It is also the story of my life because when I look back at all of the negative things that impacted my life, none of it was really my fault. They all stem out of the impact of Anti-Matter on my life's events. Could it be the same for you? Read this book and you will know.

Recent Scientific Discoveries are starting to prove that I'm right. At my website, I update all of my scientific theories.

Anti-Matter could be the key to all of our fates as individuals and even entire societies.

www.ScienceOfPhysics.com

Which was Inspired by an earlier work -

The 4 States of Consciousness

From my research that I also briefly include in this book, it is now known the Electrons possess many states, but one of them is the state of Consciousness. We know about this from the recent discovery known as Entanglement Electrons. What this taught me is that the concept that we know and love as Consciousness is not exclusive to humans, but emanates up to us through the workings of the universe.

Electrons are the first known sub-atomic region to exhibit the property of Consciousness. This book explores how they hand it up to us through the next levels of construction of all things. So, the entire universe then starts out as Consciousness as transported by the smallest of the particles that make up our universe. This energy of course cannot be trapped in such a small space - what I call the Flux and therefore it has to burst forth into everything that we see, taste, smell, touch, hear and know.

AUDIBLE BOOK

The PRINT and KINDLE VERSIONS

One of my Flagship Titles and soon to become a Classic work
of Science

The 4 States of Consciousness Inspired me to think of a way
that my theory can someday be proven.

Prior to this book, I was inspired to author - Metamorphosis -
#AICocoon. This is a deeper research into what you have read
about in this book - the Coming AI Revolution and how fast it
is coming upon us. We're about to go into the AI Cocoon. We
better make it big enough for all of us.

The Audible Book

The Print and Kindle Versions

Prior to the creation of this Book, I was inspired by my research on The Greatest Force - And How To Use it. I advise

144

anyone desiring to get the most out of this device to get this book next and absorb all of the Science. It's all briefly discussed here, but you may be the type of person who enjoys the detailed work.

THE AUDIBLE BOOK

Go To ->
http://bit.ly/greatestforce
Or search in Amazon for the Print book.

OR take the Online Udemy Course
https://www.udemy.com/starforcewars/

OR
https://www.udemy.com/4statesofconsciousness

Prior to this book, I worked in a new kind of electricity that may someday change the world. Maxtricity!

The information for this book came from the research I did for a previous book - "Maxtricity - The Science of Entangled Electrons"

THE AUDIBLE BOOK

Print and Kindle Versions

The facts will show that Electrons know about each other in certain conditions known as 'Entanglements'. This made me realize that the smallest particles of the universe - that we even use to create thought in our BRAINS have CONSCIOUSNESS and then I realized upon further study precisely how this UNI-VERSAL CONSCIOUSNESS works its way up to US.

Just after publishing 'Maxtricity', I challenged the scientists at the Large Hadron Collider to conduct an experiment to prove me right or wrong. So far, they have yet to try it. Let's give them some help.

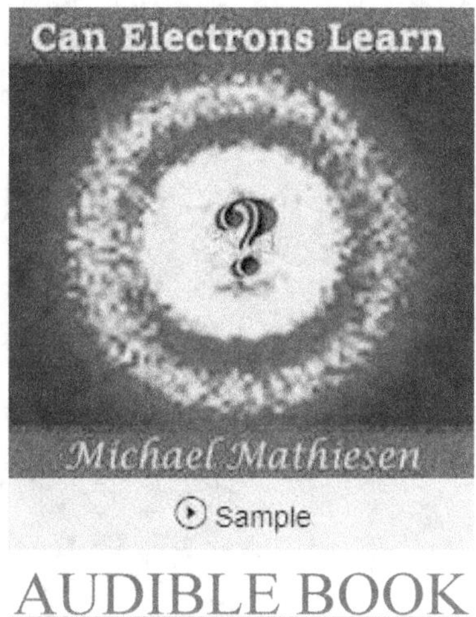

AUDIBLE BOOK

Print and Kindle Versions

OR take the Online Udemy Course
https://www.udemy.com/can-electrons-learn/

All of my books are in a line of procession from my earliest understanding of the evolution of Consciousness, to the one that I know now. It all started with the discovery of the God Particle. This is the particle that is the CREATOR of all OTHER sub-atomic particles in the known universe. So, how can we miss

the major signal that we're getting from Outer Space, the originating FORCE of EVERYTHING -

'The Force' in Star Wars perhaps?

As soon as I could comprehend this major announcement by
the folks at the Large Hadron Collider run under CERN in
Geneva, Switzerland, I wrote - The God Particle Bible

THE ONE THAT INSPIRED ALL MY SCIENTIFIC IN-
QUIRY . . .

The Audible Book

The Print and Kindle Versions

When the scientists using the Large Hadron Collider in
Geneva, Switzerland announced the discovery of the God Parti-

cle or the Higgs Boson, I was watching and waiting to get all of their data so that I could write a book that would do this discovery justice. I think I have accomplished that goal in this book.

What better way for God to create the universe, in other words, than to send out uncounted legions of his particles to create the world as they spread their influence all over Creation?

At or about this time, I also began to think about the process of human Evolution and I realized that we take our original steps just a few minutes after The Big Bang. Over billions of years, we reach a certain level of awareness that we know today.

But, we have a long way to go.

The Audible Book

The Print and Kindle Versions

All of our scientific knowledge about our own evolution starts at the formation of life on the Earth around 5 billion years ago. My work focuses on our evolution from the day that the universe was created in the Big Bang - some 13.7 billion years ago.

OR take the Online Udemy Course
https://www.udemy.com/course/thebigbangtheory/

I'm most proud of my 4 States of Consciousness because the Science cannot be dis-proven by anyone and may someday even be proven with more research into the discovery of Entangled Electrons. These are electrons and other sub-atomic particles that actually know about and react to the existence of their life-long partners, their soul-mates. I believe that the human concept of soul-mates derives from this earliest example of such a thing.

<u>TAKE THE COURSE IN CONSCIOUSNESS</u>

http://www.Udemy.com/4StatesOfConsciousness

~ ~ ~

ALSO - You can go to AUDIBLE.COM and have any of my books for Free with an AUDIBLE reading of the complete book by myself. They are Free if you simply open a Free 30-Day Trial in an Audible Account. Doing so would make me very happy because it's through my Audible books that I feel a connection with my audience and hopefully you will share this Vibrational State with me.

Prior to all of these books, I wrote:

America 2.0

Politicians keep saying that the United States is a Democracy - THAT'S A LIE - America never was

any form of a REAL Democracy and never will be until We The People are LARGE and IN-CHARGE!

AUDIBLE BOOK

And now, after the Global Pandemic has changed so much - I finally hear a chorus of folks proclaiming that this is where we're headed.

Audible Books Make a GREAT GIFT - Listen to the book narrated by the author while you DRIVE to WORK or go about your daily chores.

WE NEED YOUR REVIEWS

Wherever You Purchased the Book - Please go back, search it and then leave a brief Review. Two or three sentences is fine. This helps my books find their greatest audience because the search engines love GREAT REVIEWS. (Probably how you found this book.)

Tell Everyone you know about the Autonomous Government -
Remember our WEBSITE when posting, or tweeting.
IF they don't like to read have them search for the Audible
Version at Audible.com or go to our new Online Course -

Udemy.com/course/4StatesOfConsciousness

You can help make these great changes or leave everything the
way it is. "They" will not do it for you.

 ALSO - keep up to date with new ideas, new versions, new organizations, new things YOU CAN DO to help save our beautiful and mother of us all planet Earth

Stay Up To Date -
www.MichaelMathiesen.com

Aloha